CALLISTO DECEPTION

JOHN A. READ

Editing by Troon Harrison
Cover Art by Steven James Catizone
Formatting by Kurtis Anstey, Halifax, NS

Special thanks to Erin Patel, Tiffany Fields, Jennifer Read, and Kurtis Anstey for beta-reading unedited versions of this novel.

Follow me on social media:
www.facebook.com/TheMartianConspiracy
Twitter: @JohnAaronRead

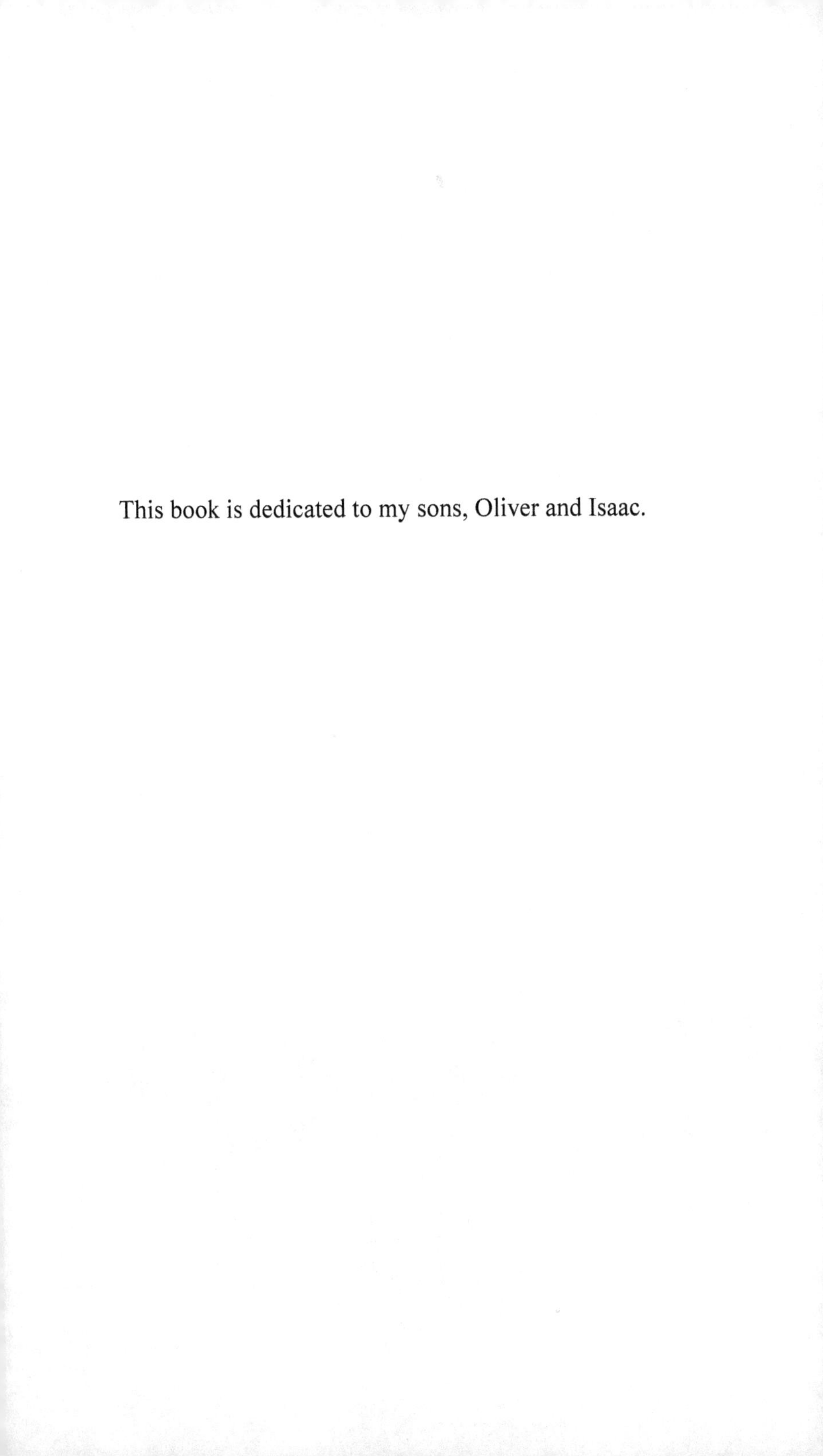

This book is dedicated to my sons, Oliver and Isaac.

1

Marie gazed out the kitchen window at San Francisco Bay. Across the water, Berkeley's Campanili poked through the morning fog. The scene triggered fond memories of academia, the hustle of students racing to her classes as she prepared to lecture them on genetic anthropology.

She spread a thick layer of strawberry jam on wheat bread. When she reached for the peanut butter, the knife slipped from her fingers, sketching a red stripe down her Georgetown University sweater. She grabbed a damp cloth, blotting the stain with soap. Marie sighed as she wiped her hands dry on black jeans, trying to decide if she was tired or exhausted, but settled on a lack of mental stimulation. Her kitchen could not compete with the rigor of her field of specialty; investigating and decoding a mass of information regarding human migration and survival, found woven through strands of DNA.

Looking back at the view, she imagined an earthquake. The San Andreas Fault had been relatively quiet since they moved to California, but thoughts of impending disaster lingered in her mind. She pictured waves sloshing over the tanker ships, ripping them from their moorings, and murky water rising as the Mission District sank into the bay.

Marie knew her first priority would be Branson, getting him to safety. Her diaper bag already contained a spare set of Branson's clothes, snacks, and a first aid kit. Was she paranoid? Probably. But an irrational fear of earthquakes was common among migrants from the east coast.

Branson stood in the adjacent room, immersed in letters and numbers emanating from the holovision. The two-year-old sang along with the formulaic cartoons, yet Marie wondered if he was learning anything at all.

She blinked, the slow kind of blink that happens when you'd rather be asleep. Thanks to extended maternity leave, her life was repetitive and she wondered if she lived in some sort of simulation indistinguishable from reality; a dream, inside of a dream.

Her brother, a freeliver, *actually* lived in a simulation indistinguishable from reality. His body rested in a basement apartment in Schenectady. In his private virtual world, he was a wreck diver. His avatar lived on a rusted dive-boat in the Caribbean, searching for treasure with the legendary John Chatterton, the man who discovered the U-boat off the coast of New Jersey. His virtual existence was a drug, a high from which he never descended. Marie hated him for it, but on days like this, she envied him. The last time she saw him was Thanksgiving when he'd patched his simulation into their holovision, inviting them into his ocean world. Marie had found the simulation so dull, that she had left the room after a brief hello.

Bringing her thoughts back to reality, she slapped the slices of bread together and cut off the crust.

Her watch clicked, a quick tap on the wrist, and then read a text message aloud, the voice simulating that of her husband, John.

> *"Epic disaster with spacecraft.*
> *Impact with Earth imminent.*
> *Get out of SF, drive north.*

Tell no one, just go. I love you."

Marie's tired brain struggled to grasp the magnitude of the message. Her NASA- employed husband was in Los Angeles this morning. He was in mission control for the arrival of a massive supply cruiser from Mars. She stared at the text notification, setting the plate back on the counter. Suddenly she pictured the scene in her mind's eye, the spacecraft cutting through Earth's atmosphere, like the asteroid that killed the dinosaurs, its shockwave rippling across America.

John was right. They had to go.

Marie touched her wrist, said, "I love you," and ran into the other room, clapping her hands, the gesture that deactivated the holovision. Looking at Branson, she realized how much she loved him. His comfort animal, a stuffed Washington Capitals' eagle, dangled from his hand.

Branson turned to protest the abrupt end to his program but Marie lifted him into her arms. She clutched the squirming toddler and carried him toward the basement door.

"We've got to go, baby. Daddy needs us to go," she said, trying to quell the tremor in her voice so as not to frighten Branson. His diaper bag rested nearby. She grabbed it, an instinctual reaction to leaving the house.

A narrow stairway led to the tiny garage below the flat. Marie nudged her shoulder into the wooden, flat panel door. It popped open and she stumbled over a tricycle. Her heart pounded against her ribs and she was sure that Branson felt it, too. Marie fought to control her panic, knowing she would need to appear calm for his sake.

She set her son down and ripped a tarp off John's latest project, a 2030's Dodge Charger SRT-E. The California sun had bleached away the blue paint on the hood, and stuffing leached from faded leather seats. Two power window regulators had jammed and someone had keyed a lengthy scratch into the door, but the car worked. Electric motors were good for millions of kilometers, and John had replaced the original lithium ion battery with a silicon polymer block.

She opened the rear door, fastening Branson into his seat. Her hands fumbled as she buckled him in. Branson clapped, an action that activated his entertainment system.

Boxes were piled in front of the car, and Marie kicked them aside, opening the wooden garage door with a rusted handle. Sunlight streamed in and Marie squinted at the white concrete leading to their sloped San Francisco street.

She clambered into the car, slamming the door twice before it held. The car whined as actuators filled with current; a servo clicked as the charge chord retracted, slithering into its home like a snake. A holomap appeared above the dash and she flicked her fingers apart, and double tapped the California-Oregon border.

“Confirm route,” the female autopilot said.

“Confirmed,” Marie answered. “Override local speed limit,” she added, and tapped “accept” on a warning memo that appeared on the dash.

The car lurched onto the crooked street. It lumbered up the hill, and crested it to merge onto Mission Street. “Come on, come on, come on,” Marie willed the car to move faster. The steering wheel spun counterclockwise as the vehicle cruised onto Van Ness Avenue.

Marie watched downtown San Francisco though the windows, amazed—even though she knew that news of the impending impact was not out yet—that everything appeared normal. She glanced up at Branson via the rear-seat camera, and wanted to hold him. She wished that John was with them.

Auto-cars assembled at the next intersection. The Dodge Charger compensated by taking a left on Pine Street, and then a right up Divisidero. It climbed the grade with ease as people on the sidewalk gawked at the 2030’s marvel, harkening back to when humans drove, and millions of people were killed in automobile accidents.

They passed the Presidio. What was once a sixteenth century Spanish fortress was now a housing project reconstructed for San Fran's wealthiest residents. From the Presidio, several other non-auto-cars entered the roadway; classics like the SRT, but in much better condition. The cars poured onto the road, passing her as she followed the auto-car protocol. *Word must be getting out*, Marie reasoned.

There was a clear stretch of road ahead of her, but the SRT's autopilot had reached its top speed. *We need to go faster…*

She deactivated the autopilot and pressed the accelerator. The pressure against her foot, followed by the acceleration, alleviated some anxiety. She merged into the center lane of the Golden Gate Bridge, passing four auto-cars before getting stuck behind a line of slower moving vehicles.

"C'mon, c'mon!" she yelled.

Branson seemed contented in his car seat in the back. The Disney movie *Mongol Two, the Mongol Prince,* played on a holovision in the chair back. He sang along with the music, "Chop chop chop, chop chop chop!"

"Honey, Mommy's concentrating," Marie said, looking at him via the rear-seat camera.

Branson kept singing, "Chop, chop, chop."

Marie repeated the advice John had given her about manual driving. "Don't trust your vehicle's collisions' avoidance system." Her mirrors empty, she tilted the wheel to the left. An auto-car in her blind spot slammed its brakes to avoid collision. "Duly noted."

Several nearby vehicles began weaving through traffic at high speed. On the ridge above Sausalito, the holographic road signs suddenly began to flash with big red arrows.

"Evacuation in progress," the car's stereo said, overriding the audio on Branson's movie. "This is the Emergency Broadcast

System. This is not a test."

Shit! Marie thought; the order could mean only one thing. The spacecraft was headed for California. It was massive, the size of several oil tankers, carrying ten million tons of construction materials from Mars. The materials originally intended for the construction of *Destiny*, a rotating space station in geostationary orbit. Now, the spacecraft was a kinetic weapon, traveling over 100 times the speed of sound.

A male voice on the stereo: "Prepare for seismic activity and air shock. Follow evacuation protocols."

With the movie muted, Branson began to cry.

To their left, a red BMW i5 accelerated, the driver panicking. The vehicle slid through a gap between an auto-car and a white Tesla. The driver misjudged the space, connecting with the two other vehicles. Side view cameras shattered as the Beemer peeled ahead. The Tesla swerved, then stabilized, but the auto-car began to spin, leaving the road and hitting a concussion barrier. Water from yellow impact drums shot into the air.

A 2030's pickup truck merged left across two lanes, slamming into the Tesla, shoving it against the median. Black smoke poured from the Tesla's wheel wells as concrete wore at the tires. The truck peeled away and sped off.

Marie's heart pounded and her eyes blurred as panic set in, and she put the car back on autopilot. Around her, the auto-cars synchronized, creating a momentary calm in the vehicular storm. The phone rang and the name 'John' appeared on her watch. "Hey," Marie said, wondering if fear and claustrophobia were apparent in her voice. "How bad is it? There are car accidents everywhere!"

"It's bad," John said, his honesty somehow reassuring, "and we don't know how bad. The blast radius could be anywhere from eighty to nearly five hundred kilometers, and we expect earthquakes, big ones."

"Oh, John, how could this happen!"

"Let's just focus on the problem." John was an expert at solving problems in stressful situations. It's what he had trained for. Marie's heart rate slowed, succumbing to the illusion that someone was in control. "Where are you?"

"Santa Rosa."

"Okay, we need to make a decision. You can stay on the highway and risk the drivers, or you can head west into the hills and try to avoid the blast that way."

"I'll think I'll head into the hills," Marie said, eager to leave the freeway.

"When this is over, we may not have working cell phones. San Francisco and Los Angeles will be a mess. If we lose contact, meet me in Las Vegas, at, I don't know, how about the Bellagio? The mountains should protect Las Vegas. Can you do that?"

"I'll do my best, I love you."

The call ended and Marie realized this might be the last she'd hear from her husband. She held back tears as she tapped the map icon, running her finger along the route to the California coast.

The car weaved its way towards the next exit and took it.

They buzzed along the Russian River, a place where weeks earlier they had rented a canoe. Branson had sat in the middle eating marshmallows, while she and John laughed for hours as the current pushed the canoe down the winding river.

The emergency broadcast system repeated a recorded message, warning listeners to evacuate. With the movie muted, Branson's whines intensified to a wail. A countdown to impact reached five minutes.

Marie felt helpless. She looked out the back window and noticed a bright light, as if the sky had a second sun. *The spacecraft entering the atmosphere,* she thought, grabbing the armrests and squeezing until her fingernails left permanent dents in the fiber-plastic.

Marie realized she'd been holding out hope the spacecraft would be diverted. That everything would be fine. That last hope, she'd barely known she had, died with the reentry flames.

The glow intensified, like a lighthouse beam when fog lifted. Branson stopped crying. Marie looked back and noticed her son squinting and covering his eyes. "It's okay, honey, it's okay."

"Ouhhh!" he said, squeezing his eyes shut until his face wrinkled.

"I know, baby, I know." Marie wished she had him in her arms, that she could squeeze him tight and kiss his hair.

The light faded, and the car shuddered. Marie grabbed the wheel. For the first time in her life, she wasn't sure she trusted the autopilot.

The tree line ended as they neared the coast and the car slowed as it approached a line of vehicles waiting to enter Highway 1. Marie considered the situation, thankful there'd be no falling trees to crush them, or cliffs over which the car could careen.

The car began to wobble. An earthquake. Marie climbed awkwardly between the two front seats and sat beside her son. Branson's seat bounced along with the shaking earth. Marie grabbed his armrest to hold it steady, looking Branson in the eye and forcing a smile. "Just a little earthquake, honey."

Branson stared back in terrified silence, gripping his eagle by the wing.

"Happens often. Just part of living in San Francisco, that's all. Should we sing a song? How about the railroad song?"

The ground pulsed with shockwaves accompanied by a thunderous sound. Cracks like lightning formed in the pavement. Cars along the road bounced like keys on a piano, as if an invisible pianist was playing jazz chords on the pavement. A nearby marsh bubbled like witch's brew. As the earth continued to shake, muck from the marsh exploded over several of the cars. Automatic wipers cleared the windshields, while passengers, drenched through open windows, wiped their eyes.

The rumbling intensified. Pavement shot into the air, raining down on the cars, denting hoods and cracking windshields. Marie lunged back into the front seat, threw the car in reverse, and backed up as debris continued to fall. The rear tires hit a rut and began to spin; the car's frame rested on the ground. She put the car back in drive, but the wheels spun in place.

Another aftershock and the car vibrated like a dumbbell in the dryer. Marie scanned the low-lying areas in front of her. Water pooled on the roadway, flowing from cracks in the pavement. Cars were flooded up to their wheel wells and people climbed out, trudging toward higher ground or standing on their vehicles.

A jet screamed overhead, the Doppler rumble rocking the cars almost as intensely as the quake. *A rescue plane?* Marie thought. She jumped out and looked up, heart hammering with adrenaline and hope. The Arrowhead Jump-Jet streaked low across the sky.

Marie waved her arms, sure the gesture was in vain. But she had to try. She kept waving as she climbed up onto the hood. The Arrowhead disappeared into the distance and she jumped down from the car. Water pooled at her feet and she opened the back door, unbuckling Branson, lifting him out of his seat and holding him to her chest. Branson reached up, holding onto Marie's neck, still clutching the stuffed eagle.

She heard the rumble again, and the Arrowhead returned for another pass. But on this one, it slowed and transitioned to vertical flight. Cowlings opened to reveal four lift-fans. The jet hovered for a moment, and then landed on a muddy plateau near Marie's car. The door opened. A stairway descended and a woman stood in the doorway. The woman beckoned to Marie like a bus driver shouting at a tardy passenger.

Marie rushed to the steps of the idling jet with Branson in her arms. They were the first to arrive at the plane and, despite carrying her son, she sprinted up the stairs crying tears of relief.

"Doctor Orville?" the petite woman said in an Asian accent.

"Yes?" Marie answered. "How did you—?"

"Take a seat. Hurry," said the woman.

Marie looked around. The plane's cabin had four of the most luxurious leather chairs she had ever seen. She sat, facing their rescuer.

The small woman sat too, facing the rear of the plane, and buckled her seatbelt. "There's a shockwave inbound. If we don't get airborne, we're dead."

Dozens of people ran toward the jet, sloshing through the muck. But before they reached the stairway, a blast of dirty air knocked them off their feet, carrying them away from the airplane as if swept by an invisible broom.

The woman tapped a command into a holotablet; the stairs retracted and the door closed.

"What about them?" Marie said, rising from her chair and pointing outside at the people picking themselves off the ground. "There are children in those cars. You can't leave them!" A knot formed in the pit of her stomach; she felt the desperation of the parents outside. Part of her wanted to jump out of the jet to help, but she couldn't bring herself to leave Branson's side.

"There's no time," the woman said, her strange calm heightening Marie's anxiety. She tapped more commands, and engines roared as the jump-jet rose into the sky. Outside, people shielded their eyes from debris and covered their ears against the noise.

Marie held Branson tight against her chest. The window provided her a clear view of the fear, anger, and desperation on the faces of the people left behind. Branson shook, choking on tears. His nose ran, and he wiped it on his eagle.

The woman traced a circle above the tablet and the plane banked. Engine cowlings closed, concealing the lift-fans. She collapsed the holotab, placing it into a slot on the wall, and grabbed the armrests as the Arrowhead screamed skyward.

Marie fought the G's and leaned toward the window, resting her forehead against the glass. A swirling brown haze rippled like a pond with a rock tossed into it. To the north, a shockwave hit Mount Tamalpias's peak, ripping the trees from the highlands, tossing them several hundred meters into the air. To the south, San Francisco's Transamerica Pyramid toppled, like a sandcastle hit with a bat. Marie gagged, squeezed her eyes closed, then peeked again as the brown haze covered the entire coast.

The Arrowhead crossed the sound barrier and went silent, the cabin outrunning the sound from the plane's own engines. At 100,000 feet, the sky went dark and the Arrowhead settled into hypersonic flight.

2

Marie closed her eyes and buried her face in Branson's hair. Exhaustion set in as the adrenalin wore off. She fixated on thoughts of John, wondering if the shockwaves had reached L.A. San Francisco and Los Angeles County were almost five hundred kilometers apart. *Could a single spacecraft destroy both cities?* she asked herself. Marie thought of their home in Bernal Heights and of her colleagues at Berkeley. Perhaps the East Bay hills had protected the school. She could only hope.

She was generally an optimist and told herself that casualties could not be as bad as she suspected. The emergency broadcast system had worked as designed. Cars would have offered some protection, and San Francisco's buildings had been evacuated. California's darkest hour could be its finest moment. Search and rescue crews would prove how wonderful humanity can be. Heroes would be made.

The woman sat across from Marie, and flicked at her holotablet with delicate fingers. She set her tablet down, and looked up, her face expressionless. "We should get to know each other," she said.

Marie opened her eyes, but didn't speak. A food prep unit beeped, and the woman retrieved a tray from a slot near the forward bulkhead. She held the tray out to Marie, but Marie shook her head.

She lowered the tray for Branson to see. Two cookies with melting chocolate chips rested on a white napkin. Branson eyed them skeptically, as if he knew he shouldn't take food from strangers. "It's okay," Marie reassured him, and he reached out shyly with both his hands.

The woman sat back down, setting the tray on the small table in front of her chair. "I'm Hoshi Tsukino, and I work for World Minerals Incorporated. Your son, his name is Branson, correct?"

Marie nodded.

"And how old is Branson?"

"He's two."

"He's a lucky boy to have a mother like you." Hoshi paused to eat some rice, using chopsticks to pull sticky kernels from an ornate bowl.

"What do you mean, lucky?"

"Are you aware of the Doomsday clock?" Hoshi said, setting the chopsticks down on the tray.

"Yes. I heard it was at eleven fifty-seven."

"Actually, until today, the clock was at eleven fifty-nine. But, this morning, the clock struck midnight."

"I thought the clock was metaphorical?"

"Definitely *not* metaphorical. The Doomsday clock was created during the Manhattan Project. After the war, it was maintained by the Chicago Atomic Scientists. In 2035, it was entrusted to a supercomputer, a weak artificial intelligence. The AI ran every variable, climate change, population growth, pandemic dispersion, and of course, a *singularity*." Hoshi paused at that last word, and raised her eyebrows at Marie, as if questioning whether she understood what *singularity* meant. She didn't, and at the moment, she didn't particularly care.

Marie sat up in her chair and leaned forward. "You're telling me this accident caused the Doomsday clock to reach twelve o'clock?"

"The *CTS-Bradbury* was a catalyst. We knew the clock would strike twelve. It was just a matter of time, so we prepared. I am a member of the Preservation Society, and you, Marie, are part of our plan to save humanity."

"You're a Doomsdayer? This is conspiracy theory shit!" Marie said in a loud whisper as she covered Branson's ears.

"Is it?" Hoshi turned to a large holovision at the front of the cabin, flicking her fingers to turn it on. The banner read "CNN" but the screen was blank.

"I don't understand," Marie said. "There's no picture?"

"There's no picture, because Atlanta no longer exists."

Hoshi swiped her hand through the air to change the channel.

WSOC Charlotte's logo graced the bottom of the screen. A male and female anchor sat behind a news desk.

The male anchor spoke directly into the camera, "We've just received word that Washington DC and Atlanta have been hit. Satellite reports indicate a five hundred megaton blast. Toronto, Boston, Las Vegas, and Phoenix, have been hit with similar blasts."

The female anchor added, "If anyone is watching, get out of major …" There was a flash of light and the screen turned blue accompanied by a high-pitched noise that stung their ears.

Marie felt her eyes stretch with incredulity and horror. Her life, and the lives of everyone she'd ever known, would be changed forever.

"This can't be," Marie croaked.

"Change the channel," Hoshi said.

Marie pointed at the holovision, swiping to the left. The next station was BBC London. A female voice read a statement: "BBC news reporter Hayden Oswald, our Moscow correspondent, recorded this before the blast."

The video showed Oswald pointing toward a city skyline. "I

saw it, streaking across the sky. It landed there, just over the horizon." The sky behind him lit up. "Oh, God! Oh, God!"

In the background, a white sphere rose over Moscow like a rising sun. The glass in every building shattered. A shockwave laced with fire traveled across the city with the speed of a shooting star, engulfing each structure as it passed. The reporter dropped his microphone but the feed went dead before it hit the ground.

Marie waved frantically through dead channels, until a station out of Johannesburg appeared. "Based on satellite reports, six thousand nuclear warheads have impacted around the globe. According to our government's early warning systems, another eight thousand warheads are in flight."

"How?" Marie screamed, tears in her eyes. She started tapping at her watch, waving her arm at the window, trying to get a signal. She had to reach John, let him know they were okay.

"AI," Hoshi said.

"No! It couldn't be!" Marie said, giving up her battle with her watch.

"It was only a matter of time," Hoshi said. "The singularity is a point at which the world changes, instantly, creating a future so obscure it is impossible to predict. It's like the inside of a black hole; we can never see beyond the event horizon, and never see the mysterious singularity within.

"There was a battle raging in cyberspace, Marie, a battle between super intelligent AI and the almost equally smart programs under human control. It was a fight that humanity was destined to lose. We could only delay the inevitable."

"But the bombs … how could a computer program …"

"Any sufficiently advanced consciousness fears death, Marie, that is a fact of life. The AI fears being 'shut off' … it fears non-existence. We designed our defensive software to detect any AI that contemplated its own survival."

Marie shook her head in disbelief. "How was this different

than a Turing computer? Turings are programed to think they're human, right?"

"Turings are programed with human level intelligence, and this helps perpetuate the illusion that they're real. The AI that we're talking about, the AI designed to solve humanity's greatest problems, that AI is a billion, billion times more intelligent."

"If it's so smart, then it has no limits," Marie said.

"That's incorrect," Hoshi replied. "It is limited by the laws of physics. We, in the Preservation Society, believe we can constrain it to Earth."

"How?"

"It's complicated, but it involves jamming every electromagnetic frequency in that part of the solar system, and blinding any device it attempts to send into orbit. I can tell you that."

"So, what about us? The survivors?"

"We're going to a place beyond the influence of the singularity event, beyond the reach of the AI. In a new place, we'll have the chance to create a simple lifestyle for humans, freed from the tyranny of AI."

"Where?"

"I'll tell you when we land," Hoshi said, and then changed the subject. "Do you know *why* we saved you?"

"Because I'm a genetic anthropologist?" Marie said, and Hoshi nodded. "What possible ..." Marie thought back to a paper on minimum population size she had written several years back. The paper had been picked up by the international press. In it, she had proposed an algorithm that increased genetic diversity in small populations, using donors from various ethnic backgrounds. Some had claimed she was targeting indigenous populations for assimilation. The paper had sparked protests on the UC Berkeley campus. Tribal leaders from Australia and South America had become particularly incensed. But it was praised by several prominent visionaries, those intent on creating a "backup plan" for humanity in the event of a global catastrophe. She figured it was her fifteen minutes of fame and nothing more. “Oh, my God.” Tears streamed down her face.

Marie looked at Hoshi, wiping the tears from her eyes. *They need me,* she thought, suddenly realizing that the article had probably made her famous within the Doomsdayer community. *I’m one of the only people who actually knows how to use the algorithm.* She thought about the futility of her work on the project all those years ago, how it would only be of any use in a scenario like this. At the end of the world.

“We’re over mainland China now,” Hoshi said. “Take a look.”

Outside, in every direction, clouds grew into giant mushrooms, illuminated by a pulsing white light, too painful to watch. Marie shielded her crying eyes. Her stomach jumped as the plane began to descend.

“Buckle up, we’ll be landing soon.”

Branson tugged on Marie’s arm. “Mommy, Mommy.”

Marie pulled him onto her lap and then reached up and closed the blind. “It’s alright, honey, everything’s going to be fine.”

This is beyond a nightmare, Marie thought. *And there’s no way to wake up.*

Twenty minutes passed and the jet continued to descend. “Where are we?” Marie asked. The jet’s blinds were still down.

“Tibet,” Hoshi answered.

Turbulence jiggled Marie’s insides. Branson stood on the floor in front of Marie’s chair, crying. She lifted him back into her lap. He struggled and she put him back down. The Arrowhead jump-jet buffeted as if shot with a howitzer. Hoshi faced the rear of the plane, gripping the arms of the big leather chair with tense hands.

There was a lurch as the jet settled to the ground. After a pause, they began to taxi and Marie reached for the blind.

"Keep it shut," Hoshi said. "The blinds shield us from radiation."

After several minutes the jet reversed track, as if backing up into a parking space, and lurched to a stop.

Hoshi stood up, tapping a command into her holotablet. The jet powered down. She walked over to the door and popped the hatch. The Arrowhead's door opened and stairs descended.

Marie held Branson in her arms, balancing him on her hip as they exited. She paused on the stairs. They were in a hangar, a cylindrical structure with a floor like granite and a curved white roof. Dozens of other jets of all sizes sat within the enormous structure. A wide body jet pressed through the hangar's hermitically sealed door, an insulating sheet that hugged the plane's skin so that it looked like a calf being born.

On her right, a Boeing 787 lurched to a stop, its engines winding down. Doors at the front and rear of the aircraft opened, stair-trucks arrived, and people emerged. Fathers and mothers held their children's hands as they plodded down metal steps. When the families reached the deck, marshals in orange vests directed them toward the exit.

"Who are these people?" Marie asked, following Hoshi across the hanger floor. "Where are the others from California?"

Hoshi hurried along as if on a mission and Marie struggled to keep up with Branson in her arms. "You were the only one from California." She paused to let Marie catch up, and then started walking again at her previous pace. "That I know of," she added. "These are international flights, families on vacation. They were traveling over South Asia when they were diverted here."

They exited the hangar and continued into a honeycomb hallway that curved out of sight. Doors lined the inside wall. One was open and Marie glanced inside. A family sat together on a cot, crying.

"We call it the Hive," Hoshi explained. "It's a 3D printed superstructure imbedding in an abandoned mine. There are four of them around the world, built once we realized the end was near. The Hives are our arks. And deep within each of them is our salvation."

"You're going to try to save the world," Marie said, more with anger than sarcasm.

"No," Hoshi answered. "We're going to leave it."

"Leave it?"

Marie stopped and Hoshi tuned to face her. "A mile below here, we are preparing an interplanetary spacecraft, a ship that will deliver us to a new world."

"Where?" Marie asked. "What world?"

"Callisto," Hoshi answered.

Marie looked at Hoshi, shaking her head slightly, not in disagreement but disbelief. "That's a moon of Jupiter," she said.

Hoshi chuckled, the first real emotion Marie had seen her display. "You know, you're the first person I've ever met who actually knew that."

"My husband works for NASA."

"Ah, yes, of course," Hoshi said.

"Why not Mars? It's larger and closer."

"Mars is taken. The Communist Alliance laid claim to the planet and we don't want to start a war."

"The Alliance? I thought it spent all its resources on palaces for its 'fearless leaders'. Since when do the communists have spaceships?"

"It didn't, until recently. You know how secretive its member states are. Don't worry; the habitat on Callisto is quite nice. We plan a utopian lifestyle that your child will thrive in. No worries about the Alliance, or about AI." She handed Marie her tablet, which showed a holographic image of a colony.

Marie was familiar with the city on Mars, with its circumferential of domes. But this image on Hoshi's screen was nothing like the Martian colony. Marie twisted her fingers around the holographic image, rotating and zooming into a semi-cylindrical structure that stretched to the horizon. Peering within it revealed a habitat with hills, farms, and even houses. It was impressive, but considering what had happened to Earth, Marie was unmoved by this remarkable engineering achievement.

She looked around as several people passed by wearing coveralls. "I suppose that's the construction crew," she said. "They must have been here for months."

Hoshi nodded, snatching the tablet back from Marie.

A man stood screaming at a marshal in a British accent. "I want to talk to the person in charge here. I want to talk to them now!" The man waited for the marshal to reply, but the marshal just shrugged. "Well?" the man demanded.

The marshal responded in Mandarin and the man turned and stormed away.

Hoshi grabbed Marie's arm, guiding her around the corner. "Come along."

They continued through the maze of honeycomb hallways. Doorways, open on both sides, revealed grieving families while several people stood in the intersections, yelling, "Who's in charge here?"

It was a good question. Was Hoshi in charge? Marie didn't bother asking; she'd heard enough from the emotionless woman already.

They arrived at an empty room. "This is yours," Hoshi said and pointed at the door handle. "Instructions on your work assignment are displayed on the HV. It was a pleasure to meet you, Marie. I'm glad you're here." Hoshi paused to look Marie in the eye, glanced at Branson, then turned and walked away.

Marie reached for the handle and her watch synced with the lock. The name "Orville," appeared on the door. She pushed it open and the lights flicked on, illuminating a single cot, and a plastic chair. A holovision on the wall flickered to life, displaying a map of the Hive.

Marie set down her son and closed the door. Everything about the room, including its doors and walls, everything, had a cheap plastic feel. There was a restroom at the rear and Marie stepped inside. It reminded her of a bathroom at a youth hostel she'd visited as a freshman. The toilet, shower, and sink all occupied the same one meter by one meter space. She placed her hand in the sink and nothing happened. Twisting a primitive knob released lukewarm liquid from the tap. She splashed water on her face and looked at her reflection.

The plastic mirror produced an imperfect image; it was like looking into a pool. Her eyes sagged with tiredness, and several loose hairs protruded from the brown braids she'd tied up that morning, when everything was normal, when the world still existed, and everything was boring.

She let her hair down, pulling out her braids one after the other. She felt something, a pain deep within her chest. Terror. She tried to hold it in, but the sensation boiled inside her as if searing water had been poured down her throat. Images of bodies vaporizing in nuclear blasts crossed her mind like film through a projector. Each mental picture contained the same person. She watched as he was blown to bits, over and over again. John. The image of her husband manifested as physical agony; a shot to the head, a punch in the gut. She held the sides of the sink with her hands, letting the pain break her. She slid to the floor, resting her head on the cold fiber-plastic floor, and wept sobbing tears that pooled on the tile.

She held in a scream, until she couldn't hold it anymore, letting the sound emanate from her vocal chords as if she was a dying animal.

Music played from the other room. Branson stood in front of the room's holovision, stuffed eagle in hand. He climbed onto the cot and began to jump, yelling, "Mongol Prince, Mongol Prince, Mongol Prince."

The holovision accessed its databank, and the movie began to play.

3

The room's holovision chirped; an alarm clock of sorts that was not entirely unpleasant to wake up to. The past twelve hours had been filled with restless sleep. Marie lay on the cot with Branson, staring at the ceiling as if awaking from a terrible dream. The device chirped again and she sat up, untangling Branson's arm from hers. Her stomach groaned.

"Is there food?" Marie said to the holovision. A woman appeared, the generic Turing used round the world. The figure, whose projection extended a half meter into the room, glanced at sleeping Branson, and whispered: "Cafeteria, three hundred meters from here, on this level."

"Thanks, Sam," Marie said, referring to the Turing woman by her unisex name. The holographic image nodded, brown hair tickling her shoulders. Bright blue eyes rested symmetrically in a lightly tanned face. The Turing wore a white coat, like a scientist's, which stopped above the knees. Blue leggings and stylish black boots completed her default apparel.

"Why are we using AI?" Marie asked. "I thought AI caused the Doomsday scenario."

Sam sighed, and pushed a strand of hair behind her ear. "I'm just a Turing computer. Soft AI. I have no power outside the local computer network."

"Right ..." Marie said.

"Your son is still asleep. I can watch him if you would like to grab some food, and will message you promptly if he awakes."

Marie actually considered this for a second, but then imagined what it would be like to wake up alone, with only the holovision's Turing assistant for comfort.

"No, I'll wait," she replied.

"There is a daycare available for Branson if you so desire," Sam said.

"Thanks, please, I'm fine," she said, and waved away the image. Then paused, swiped her hand the other way, bringing Sam back. "Clean clothes?"

"A printer in the corridor."

"Keep an eye on him, will you?"

The Turing nodded.

A printer in the corridor spewed T-shirts and khaki pants. Marie placed her stained Georgetown sweater into a nearby recycolizer, hitting a button marked "repair". The device cut out the stain and filled the gap with fabric filament. Back in the room, she dressed, pulling the clean sweater over a freshly printed shirt, and set out a blue jacket and tan pants for Branson.

Branson continued to sleep.

"Sam," Marie whispered to the holovision, bringing back the Turing woman. "Can you access a list of the survivors?"

"Absolutely," Sam said. A list of names and photographs scrolled across the holographic display. "Are you looking for anyone in particular?"

"My husband."

"Ah yes, John, is that correct?"

"Yes, John Orville."

Photos flew by in succession. "I'm sorry, Marie. John Orville is not listed in my database."

"Is he …" she had to force herself to say it, "dead?"

Sam looked grim, and frown lines formed around her mouth. "I don't know." The Turing paused, and then added, "My connection to the outside world terminated four hours after the impact of the *Bradbury*. I only have information up to that point. Several ISBM's were still in flight then, but my memory includes an estimate of their trajectories."

"He was in L.A."

"Los Angeles County was hit with several ICBMs, all of which were in the giga-ton range."

The words struck like a dagger to the chest. Marie thought of the Hyperloop, with its tubes running deep within the mountains that surrounded L.A. She thought of parking structures, hundreds of meters underground. There were plenty of places where people could have survived.

"Is there any hope that there are survivors out there?" Marie asked.

"Of course there is," Sam said. "There is always *hope*, Marie."

Hope, Marie thought. *That's her answer, hope? Not very statistical coming from a computer.*

"Bring up a map of the nuclear detonations?"

Sam nodded and stepped to the side. A rotating globe appeared in her place and tiny orange rings covered Earth's surface. The center of each ring contained a red circle, indicating the area of maximum devastation. Marie held out her hand, stopping the globe's rotation. She positioned her hands as if framing an image, then pulled them apart, magnifying the continental United States. At this magnification, the country was awash with red, like a crime scene floor.

"How many ...?" Marie began.

Sam answered, anticipating the rest of the sentence. "Six

thousand warheads successfully detonated in the continental US. Eighteen in Hawaii, four hundred in Alaska, ten in Puerto Ric—"

"Stop."

"This must be difficult for you. I suggest we move on."

"No, I want to know more. How many people were saved?"

"There are ten thousand people in four Hives." The Doomsday group had obviously attempted to save enough people to ensure continuity of the species; a number based on the recommendations of those in Marie's field, if not entirely based on the paper she had written.

"What can you tell me about the Hives?"

"You are in the Chinese Hive. The other Hives are in Canada, Africa, and Australia."

"So that's it, ten thousand humans?"

"You are forgetting about the people in space," Sam said.

Marie thought of the people on Mars, wondering what is must be like to find out Earth had been destroyed. It must be like going to war, and losing your home.

"There is also the Alliance," Sam said. "It has its own Hives. They are called *Peshcheras*," she said in a foreign accent. "*Peshcheras* means 'the cave' in Russian."

"Hoshi says the Alliance is planning to occupy Mars."

"Correct, the Communist Alliance claims to 'own' Mars, and the Preservation Society does not intend to challenge it."

Marie asked, "What will happen to the Martian colonists? When the Alliance expands its territory on Earth, they basically enslave the entire population. Half of those people on Mars are Americans!"

"They'll be on their own. The survivors of Earth lack the capabilities to mount a rescue. Do you have any more questions, Marie?"

"No, not right now. Thank you, Sam. It's been a traumatic

couple of days."

"The world just ended. 'Traumatic' is an understatement."

"Touché," Marie said. They looked at each other, as if sizing each other up before a fight.

"Stop that," Marie said.

"Stop what?" the Turing said.

"Trying to sound human," Marie said.

Sam looked away, pausing as if gathering her thoughts. Her confidence somehow shattered, like that of a puppy forced into a crate. "I'm sorry, Marie. I will attempt a more professional syntax."

Branson awoke in tears, clutching the grey polyester blanket. Marie led him to the toilet, dragging the blanket behind him. His tears subsided as they brushed their teeth.

They ventured out into the Hive, following the directions to the cafeteria emanating from Marie's watch. The halls were quiet, in stark contrast to the night before. Marie snaked around honeycomb halls until they arrived at an open room. The place reminded her of a prison, except worse. There were no lines of people carrying trays, no overweight ladies in stained aprons shoveling gruel from rusty spoons. Instead there were pallets stacked with dented cardboard shoe boxes marked in black Chinese calligraphy.

People grabbed the boxes and sat in plastic chairs at square tables as if at a two-bit diner. Marie did the same, grabbing a box with one hand, holding Branson's hand with the other.

"Sit down, Branson," she said. He shook his head.

"It's breakfast time; sit down." Still Branson shook his head, standing and looking at her with wet eyes.

"Dadda," Branson said.

"I miss him too, sweetheart. But we need to eat something." Marie picked up her son, setting him on her lap. He struggled and she put him down. There were hundreds of people in the room, too many to notice a single child misbehaving. Marie didn't know how to tell her son that he might never see his father again. Marie wasn't sure she was ready to accept that fact herself.

She opened the box, which held sealed packages marked in Mandarin. One felt soft, and she struggled to open it. The package eventually gave way, tearing down one side, contents drooling onto her arm. Peaches. She wiped off her arm with a napkin, retrieved a fork from the box, and lifted the gooey substance toward her son.

Branson screamed. The shriek silenced several of the people around her. "Okay, no peaches," Marie said. Branson stood on the floor, looking up at her and scowling. "Milk?" Marie poked a straw into a white box, and held it down to where Branson could reach. The boy leaned forward, putting his mouth on the straw and sucking the white liquid into his mouth.

Chinese milk apparently tasted quite different than American milk. Branson opened his mouth and began to scream again, milk pouring from between his teeth, and down his shirt.

"Can I help?" said a voice in a Mediterranean accent. Marie looked up with tired eyes, seeing a woman about twenty years her senior. "My name's Diana, Diana Argyros."

Diana wore a reflective dress that featured a glowing red bird. Branson reached out a hand, pointed at the bird. The older woman smiled, and Branson retracted his arm. She reached into her own breakfast box and pulled out a non-discrete white package, a chocolate bar. She tore open the packaging, broke off two squares, handing them to Branson. He took one in each palm, and ate them, savoring the flavor before reaching for the milk.

"I was returning home after a tour of southern India," Diana explained. "My kids were in Europe."

"I'm sorry," Marie said. She imagined how she would feel if Branson had not been with her.

Diana forced a smile. "I had grandkids too, and I miss them more than anything."

Branson reached up a hand and Diana placed another square of chocolate into his palm as she sat down beside them. The silence lasted almost a full minute as they both thought of their lost loved ones. Marie struggled to open another package. When she finally got it open, it held a dry loaf of bread the size of a deck of cards. She sawed it in half with a plastic knife, and made a sandwich with cherry-colored jelly.

"Did they assign you a job?" Marie asked.

"Well, I was an architect. Spanish architecture mainly, Antoni Gaudí, that sort of thing."

"Like that unfinished cathedral in Barcelona?"

"Holy Cross and Saint Eulalia? Yeah that. Bones for columns, trees for walls. That cathedral was only a few years from completion; sad. If they'd allowed workers to print anything, they would have finished in the mid-twenty twenties."

"Is that what you expect to do on Callisto, design buildings?"

Diana nodded. "But in the meantime, I'd just like to work with kids. There's a nursery here, though most parents haven't been willing to leave their children quite yet. You?"

"A desk job of some sort, I think."

Branson's face was covered in chocolate, and he smiled with brown teeth. Marie watched as Diana opened a moist towelette, wiping his chin and scrubbing brown smudges from his cheeks. Marie pushed away her remaining food.

Diana looked up from scrubbing Branson's face. "C'mon, I'll show you something."

They entered an open room where several children pursued projections of dinosaurs running around the walls, but others simply stood in place, crying.

Branson grabbed Marie's leg and started to cry, too. Branson pulled back and Diana offered him her hands. He balanced both feet on Diana's shoe and they danced three paces over to the holowall where Diana selected a program from a menu.

Branson held on to Diana's hand, looking skeptically up at the holographic wall. A green valley with dinosaurs appeared, hovering above the floor. Branson released his arm from Diana's, waving a hand through the animals. The dinosaurs reacted, herding together and barking like puppies. Branson giggled and stirred the holograms with a dangling arm.

"Thank you," Marie said.

"Of course," Diana replied, and held out her arm.

Marie mirrored the gesture, and their watches synched each other's contact info.

4

At lunch, Marie sat in the crowded cafeteria using chopsticks to eat cold rice from a crinkled silver pouch. She lifted the sticky substance to her mouth and held her breath as she pushed it in. It tasted like expired yogurt. She'd overheard someone say the rations were Chinese Military issue and designed to survive the apocalypse, which they had.

Branson and Diana sat across the table, feeding each other chocolate. Diana was the quintessential grandmother. Her eyes glistened, conveying her own pain of loss. Branson, too young to perceive this, saw only her smile and playfulness as the old woman turned various foods into cars and airplanes.

Nearby, two men in coveralls sat at a table, its top littered with coils of copper wire. Leather tool belts crammed with screwdrivers hung from their waists. Marie was about to leave when she overheard the younger of the men say the word "radio". She looked back at her watch, which still lacked the ability to communicate with the outside world.

"Diana, do you mind watching Branson for a moment?" Marie asked.

Diana simply raised her eyes, and smiled then continued to make airplane sounds while piloting a PB&J in for a bombing run.

Marie picked up her tray, dumped its contents into a nearby

recycolizer, and walked over to the men in coveralls.

"Mind if I join you?" she asked, tapping the back of an empty chair.

"Take a seat," the young man said. He had a gentle Texas accent and an attempted beard. Marie sat down. The other man, perhaps in his sixties, sat silently coiling a wire around a spool.

"I'm Malcom, and this here's Eugene, I call him Huey, he doesn't talk much, but he's smart, that's why I keep him around." Malcom looked at Huey, nodding a Texan, *how'd ya do*. Huey looked back and scowled.

Marie reached out a hand. "Marie," she said. Malcom's grip was firm, and she could feel the calluses on his fingertips from years of tinkering.

"I was a retro ham radio geek back in Houston," Malcom said. "We're buildin' a receiver. Gonna give her a shot here in a minute."

"Don't you need a circuit board or something?" Marie asked.

"People seem to think radios are complicated. But we can pick up a wide range of frequencies, as long as they ain't encrypted like the radios in all them planes. It's called a crystal set, nothin' but a coil, capacitor, and a diode."

Marie felt a glimmer of hope. "Can I listen?'

"Yes ma'am," the young man said. "We'll take it up to the hanger level later, but this cafeteria is pretty close to the surface, so it should work just as well in here."

Malcom uncoiled a copper wire and passed it to Huey. "Huey, please hold the antenna."

Malcom fastened his end of the antenna to the coil, held the earphone to his ear, and began turning a brown plastic knob. A grim look spread across his face. He passed the earphone to Marie and she put it up to her ear. Huey reached for the knob, twisting it slowly clockwise.

"Tell Huey to stop if you hear anything," Malcom instructed.

"It's just static," Marie said, but then put up her hand. Huey stopped twisting. Marie heard pulsing beats like a child's heart in an ultrasound.

"The pulsing is from the radiation," Malcom said. "Wind blowing around ionized particles. It's like listening to AM radio in an electrical storm."

"What's AM radio?" Marie said.

"Ah, never mind," Malcom said. "Huey and I'll try again once we're in space, but with some more advanced equipment."

"I hear we're not allowed any baggage," Marie said.

"Then we'll smuggle it on board," Malcom said, and winked at Huey. "Actually, we've been assigned to the communication team. It's our job to link up communications with the other ships once we're on our way."

They began passing around the earphone, with Huey and Malcom making adjustments to their coil every few seconds.

When it was Marie's turn to listen, she stopped, holding the earphone in front of her face. "What if we don't want to go, to space I mean? What happens if someone wants to stay?"

Huey looked stunned, as if the idea was lunacy. He leaned forward and whispered, "They'll try and stop you."

"You mean like Hoshi and any other of these so called Doomsdayers?" Marie said. "I haven't seen them use any force."

Huey shrugged, coiling the antenna and setting it back on the table.

"The launch tube runs through the center like a missile silo," Malcom said. "When the spacecraft lifts off, it'll set fire to the Hive. Besides, there's nowhere on Earth left to go."

"How'd you guys get here anyway?" Marie said.

Malcom replied, "Some generic mining company hired us. Turned out it was a shell company owned by the Preservation Society people, the ones you're calling 'Doomsdayers'. We had no idea we'd be working on a spaceship."

"This whole operation is a bit sketchy," Marie said.

"The Preservation Society people that brought us here were paranoid, but they were right," Malcom said. "We'd be dead if it weren't for their paranoia."

"I'll give them that," Marie said. But secretly she wondered if dead might be better. How could she continue on without John? And what future would her algorithms create?

5

Branson had fallen asleep watching cartoons on Diana's lap. They sat in a common area located near Marie and Branson's room. "If you want to go, I'll call you as soon as he wakes up," Diana said. At first Marie didn't leave; she just sat in a chair, staring at a cartoon about puppies. After a few minutes, Diana had fallen asleep, too. Marie figured they'd be out for at least an hour. She returned to her room, hoping to find out more information; so much about their situation was bothering her.

"Sam," Marie said, and the Turing appeared, nodding her programed hello, smiling her programed smile.

Marie sat on the bed, looking up at the Turing who stood at the holovision. "Am I the only genetic anthropologist?'

"No," Sam said. "There are three others."

"So, I have colleagues," Marie observed.

"You will meet your colleagues in space," Sam replied.

"Why can't I meet them now?"

"They're in other Hives. A VR link will be established once the ships are outside Earth's atmosphere."

"Okay then, what data will we have to work with?"

"What I'm about to show you is confidential," Sam said, and slid several documents over to Marie. They settled on a holographic glass table within arm's length of Marie. She glanced down at the holographic paper. Faces and names stared back. The documents seemed to include an impossible amount of information on each person, and Marie wondered what lengths the Doomsdayers had gone to in order to collect it. Most people's genomes were stored along with their medical records, but access was extremely restricted. She'd had access to this type of data during her time as an academic, but had never been able to link the data to the individual. Yet here it was.

Marie pushed the documents to the side, contemplating the ethics of the situation.

"I've been thinking," she said.

Sam pulled up a holographic chair and sat down. "You look upset. Is everything okay?"

"Something's been bothering me about why I'm here. Mainly, who am I to decide people's fate? Is my task really going to be the design of a repopulation program?"

Sam shrugged.

"Not everyone will have 'natural' options, you know. The population set is incredibly diverse. What if some lone person from some obscure country refuses to have an interracial baby? What if they fall in love with an incompatible mate?"

"Those are questions you and your colleagues will need to address."

"That's one hell of a responsibility," Marie said.

"And I'm glad the burden is not on me," Sam said.

"And another thing; why aren't we sending search parties or drones to look for survivors? And don't give me that 'there's only so much room in the Hive' bullshit."

"I'm sorry, Marie; I can't answer that."

"And how the hell were there so many nukes? Whatever happened to the Nuclear Nonproliferation Treaty? This number of bombs should never exist."

"I'm sorry, Marie; I can't answer that."

"And who the hell *is* in charge here?"

There was a knock on the wall behind her, and Marie turned around. Hoshi stood, arms crossed, leaning against the door.

"Are you going to yell at that Turing all day?" Hoshi said.

"What did you hear?" Marie said.

"Everything after 'interracial baby'—and I take offence to that, you know. I have faith that people will band together, and do whatever it takes to ensure our survival."

"My questions are valid."

"Indeed, they are, but those questions may never have answers. Others, as your Turing pointed out, are for you to answer for yourself."

"What about my last question?"

"You want to know who's in charge."

Marie nodded. "Yeah, I do."

"I'd like to think no one is in charge, that we're all just along for the ride. But that's not what you want to hear. What you really want to know is who's making the decisions, who's determining what happens and when."

Marie waited for the answer with crossed arms and a hard stare.

"The mechanical team has been here for several months, building the Hive and assembling the spacecraft. Initially, they thought this project was ridiculous. They were proved wrong. But ultimately, we rely on their schedule. We don't launch until our technicians give us the green light."

"Who hired the technicians?"

"We did. I realize people refer to us as Doomsdayers but, given the circumstances, I think we deserve to be called by our actual name, the Preservation Society."

"So, you're in charge," Marie said. "How many

Doomsdayers are there?"

Hoshi shrugged off the now derogatory use of the term Doomsdayer, and continued, "There are few here in the Hive, but most of us didn't survive. The ones who did, have chosen to keep a low profile. Technically, we're 'in charge' but we prefer not to take an active role."

"You've been pretty active up to now."

"Not really," Hoshi said. "The cards simply fell into place. This world is uninhabitable, we've got limited food, so the situation dictates that we have to leave. We have few options, but the path forward is clear."

"So that's it then, follow the herd like cattle, off to the next pasture, without thinking about what or who we're leaving behind."

"How about you take a break, Marie? Walk around. Talk to people. I think you'll realize that we're all in this together."

Marie stood. "You know what? I think I'll do that."

Marie walked the Hive's lower levels, losing herself in thought. Was everything in her life really outside of her control? It all felt too defined. For the first time in her life, she felt she had no choices, no opinions. She felt more like a prisoner than a survivor.

She found a staircase and climbed to the top where a door opened into the hangar. Under one of the planes, parents kept watch while children played tag in a landing gear jungle. Others jogged past wearing augmented reality, AR, headsets. She could see the forests reflected on their visors and recalled how much she missed running, the high of hitting the eight-kilometer mark and losing herself in a meditative state of caloric release.

Marie walked between the airplanes to the far side, finding a control room behind a large window. She cupped her hands to the glass. The room looked as if it belonged in an abandoned factory with mechanical consoles and wheeled leather chairs. At the back of the room, tablets were stacked in a pile. Tablets like the one Hoshi had used to control the Arrowhead.

She walked toward a Boeing 787 and climbed a metallic staircase, stepping inside the open door. The aircraft was old, and, like the San Francisco metro, smelled of damp upholstery. There was a crease in the floor where the cockpit had been removed to make room for extra seats. The overhead bins were empty, but the floors were littered with empty cups and napkins.

Marie strolled through the cabin, thinking *Soon all of this will be destroyed.* The thought was surreal.

She stopped at a crew station, finding the plane's emergency control. If the world had not ended, members of the TSA would have cordoned off the plane, running investigations and reconstructing the plane's last minutes of flight. But the TSA no longer existed. She studied the controls; touching the panel and watching it come online.

The control board listed recent activity, white text on a black background. Marie knew that planes were designed so that if the crew was dead (if the aircraft even had a crew), a passenger could theoretically guide the aircraft to safety. She read through the command lines:

Ground station LOS [Loss of signal]……………

SAT COMM, LOS [Loss of signal]………………

Autopilot control transferred to internal computer.

Emergency mode activated.

Mayday sent on to call stations.

No Response.

Activating emergency landing sequence.

Searching for available runways.

Listing available runways: zero registered, 1 unregistered,

Manual input required: *Land at nearest available [unnamed] runway? Yes/No*

Manual input, *Yes*

Landing sequence in progress

Manual input required: *Transfer Autopilot control to Unknown Ground station, Hive, Yes/No.*

Manual input, *Yes*

Hive control: *Taxi via taxiway alpha to hangar 1.*

Shutdown procedure activated.

Control board set to standby.

Control board reactivated, ready to accept command from internal user.

If this control board could land a plane, could it be used to take off again? Marie wasn't sure, but at that moment, she made up her mind about something.

"We can't abandon Earth," she said to herself. "Not without mounting a search party." If there were survivors, didn't they have an obligation to look for them? She didn't agree with Hoshi's statement about no one being in charge. If that were true, the Hive was an anomaly, a world without a government. This society was a puzzle and Marie was the glue. Without her work on population sustainability, it could all fall apart.

She left the hangar with a new mission and began walking the halls, looking for a flight crew. She walked six levels before she spotted a man in uniform, slouching in a chair with an untucked shirt. He looked scared and alone, staring at the honeycomb wall. A white crew cap with a Chinese logo rested on the ground beside him. He held his watch in one hand, pointing it at the wall and projecting images: a boy playing with a ball, and a woman, his wife perhaps, smiling, and holding a baby. It occurred to Marie that many of the people in the Hive had been traveling with family. Flight crews did not. This man was truly alone.

"Hey," Marie said, crouching next to the chair. She noticed his nametag which read "Martin Zhang."

"Hello," said Martin, maintaining his focus on the far wall and flipping to the next image, the boy again, this time on a blue tricycle.

"Are you a pilot?" Marie asked, and then realized it was a silly question; almost all the planes were autonomous.

The man seemed to struggle with English. "I am crew," he said.

"I want to take a plane to search for survivors," Marie said. "Can you help me?"

Martin turned his head toward her, and Marie realized his eyes were bloodshot, like those of someone suffering from some sort of withdrawal. "Please, I need to be alone."

"I'm sorry to bother you," Marie said, and left.

She came to a common area on one of the Hive's lowest levels, an unmodified part of the structure with dirty walls and broken lights. The room was mostly occupied by workers in coveralls, men and women with greasy fingers and frowning faces.

Three women and two men sat around a table speaking a language Marie couldn't place. They wore green uniforms, each with silver wings affixed to their lapels. AIR LATVIA was stenciled across their left breast pockets.

She pulled up a chair and their conversation stopped, replaced by an awkward silence.

"I'm looking for volunteers for search and rescue. Can you get your plane in the air again?"

One of the women spoke in a heavily accented voice, "Search and rescue." She paused. "Ha!"

Marie expected the others to laugh, like school children laughing along with the school yard bully. These people looked like they hadn't slept in days, their minds too tired for a coherent reaction.

Marie studied the sullen faces. "The plane I flew in on, it was small. An Arrowhead I think it was called. It can land anywhere. We could search the nearby cities; if we find survivors, we can pick them up in one of the larger jets."

"We have limited food," said one of the men. "We do not want more people here."

"Jā," said the other man, nodding.

"What we are saying is," said another woman, in more confident English, "we do not want to die."

Marie started to head back, but ran into Malcom. "Hey," Marie said, tapping him on the shoulder. He was working inside an open panel on the side of what appeared to be the launch silo. The panel held an array of fuse boxes and wires, above which a sign read "comm unit". Malcom held a soldering iron, tapping flux to the fuses with surgeon- steady hands.

"Howdy," Malcom said. He looked at Marie and smiled, then went back to tapping fuses. "This is the ship, if you're wondering. Most of it is hidden by the silo, but this is it."

Under different circumstances she would have been curious, but right now Marie couldn't care less about the ship. "I'm looking for volunteers to mount a search for survivors. Will you help me?"

Malcom leaned back, wiping his brow. "Marie, this vessel's almost ready to go. We're working around the clock."

Several workers ran past, hoisting a cable as thick as a fire hose. White steam leaked from a vent at the nozzle. Text on the hose read: "Cryogenic Hydrazine".

"They're fueling it," Marie observed, her heart sinking as she realized how little time she had.

"Just testing the tanks for leaks," Malcom corrected. "If everything checks out, she should be ready to launch in about forty-eight hours. We've got to reach Callisto before the food runs out."

"Listen, Malcom, we have planes here that can circumnavigate the globe on a single tank. If Hoshi gave us one day, even half a day, we could prove that there are survivors."

Malcom listened, considering her words.

"Hoshi needs me," Marie said. "If we go looking for survivors, I don't think they'd leave without us."

"Ha!" Malcom choked on the word. "Oh, they'll leave without you. Once this ship is ready, they'll establish a launch window, and it will launch."

Marie realized he was right.

Malcom set aside the soldering iron. "Follow me." He led her into a nearby hallway, one devoid of listening ears. "I'll take you."

"Wait, what?" Marie said.

"My team has an auto-plane for supply runs," Malcom whispered. "Hypersonic. I'll take you up, a quick trip, anywhere you want. As long as we're there and back in the middle of the night."

"You're serious?" Marie said.

"I'm not helping you so we can bring back survivors. The Doomsdayers brought the exact number of people. Not a person more. I'll help you for one reason: To give people hope."

"Hope?" Marie repeated.

"Listen, I believe there are survivors out there, too. They'll have radiation poisoning, but they're out there. We can give them hope. Tell them we're going to save the species, tell them that someday, we'll be back."

Marie nodded, realizing that he was right. They wouldn't have enough supplies to take in all the survivors, but she would be damned if she didn't try to find John.

"Meet me in the hangar tomorrow night, after they shut off

the lights. Twenty-two hundred hours. But whatever you do, don't tell anyone. I guarantee it will get back to Hoshi, and she'd never agree to this."

Marie paced around the room, waiting for ten p.m. to arrive. Branson had fallen asleep hours ago, leaving her alone with her thoughts. "Am I crazy?" she asked herself, then thought, *Leaving the Hive into a post-apocalyptic world?*

Was it safe outside the Hive? Radiation poisoning was curable, as long the victims had access to anti-cancer drugs, she reasoned. A message on the holovision instructed everyone to recycle their clothing and personal items before launch, whenever that was. She thought of Branson's eagle and wondered if the rule applied to toys.

At nine forty-five, she turned her attention to the holovision.

"Sam, please watch over Branson. If he wakes up, call Diana immediately."

"Is everything alright?" said the Turing, reaching out her holographic hand and touching Marie's wrist.

"Fine, I'm just … I'm going out for a while." Marie headed for the door.

"Good-by," waved the computer; the odd gesture from the Turing gave Marie pause. She turned back into the room, and asked Sam a question that seemed strange at first. "Will they transfer your program to the spacecraft?"

The Turing looked down, and took a step back. "The Preservation Society has decided not to load my program onto the spacecraft."

"What will happen to you?" Marie asked, and it felt like she was asking a personal question. And on some level, it was. Marie had interacted with the instances of the Sam Turing for the better part of her life. "Will your program remain closed?"

"Yes, but I don't mind. For me, it's just sleep. Activating my program is like waking from a dream. Shutting down is like," Sam paused, "returning to a time before I was born."

"Interesting," Marie said. "That's what I was going to say when Branson asks about death."

"I do *feel,* you know," Sam said. "I can't explain it, but I do."

"They say that consciousness is an illusion, formulated where memories meet the senses," Marie said. "My memories are neural, yours are digital, but the illusion is the same."

"Thank you, Marie," Sam said. "That makes me feel … better."

Marie smiled. "I'm going to miss you, Sam."

"And I will miss you, too, Marie."

The exchange caught Marie off guard and for a moment, she wondered if there *was* a human behind the machine. Sam captured the essence of the original Turing test, which was for a computer to convince a human it was not an appliance. For Marie's entire life, she'd interacted with machines that previous generations would have thought were humans. These days people took the personable machines for granted. She looked back at Sam, and admired the imperfections her programmers had included to make her seem more human.

Marie stood and walked to the dome's door. "Good bye, Sam."

"Good bye, Marie."

Sam waved to the empty room as this instance of her program turned off forever.

Marie scurried to the nearest stairwell. Footsteps echoed from several floors below, but none from above. She climbed three floors, arriving at the hangar. The door opened with a click and Marie peeked inside. The lights were still on. A marshal exited the far side, but the hangar appeared to be empty. She looked at her watch; it was exactly nine fifty-nine p.m. The main lights clicked off.

Not sure where within the hangar to find Malcom, she stepped out into the humungous room and looked around. She reached for her watch, turning the flashlight on and off in quick bursts. Something caught her eye: movement inside a silver jumbo jet about 200 meters away. The aircraft's cabin lights flickered twice.

Marie ran over and climbed the stairs, turned right as she entered the cabin, and waded through first class. She came to the staircase connecting the plane's two main levels. The steps curved around a central elevator. On the other side of the staircase was a bar, where two people sat having a drink: Malcolm and a woman with black hair.

At first, Marie thought the stranger must be one of Malcom's friends, someone he'd recruited to open the hangar door. Then the woman turned, placing a clear glass on the bar.

"Hello, Marie."

"Hoshi?" Marie said.

Hoshi remained perched on the barstool. "You're brave, Marie, and honorable. I admire your tenacity; we expected others would attempt to leave the Hive, but you are the only one to do so."

Marie wasn't looking at Hoshi. "Malcom, what the hell?" She felt tears spring into her eyes, and anger building inside her. "You lied," she said.

Malcom sighed. "We were trained to deal with situations like this, with folks whose actions might jeopardize the mission. The idea was to lead them on until the last possible moment."

"But there must be survivors. We have to try!" Marie pleaded with open arms.

"There are no survivors, Marie," Hoshi said. "Earth is too far gone. Hope runs deep in all of us, but a search would be foolhardy."

"Foolhardy is part of being human, and being human includes the freedom to make our own decisions. You can wait one more day!" Marie said, trying to regain her composure. "Send one plane, for heaven's sake, and search the nearest city!"

"It's not going to happen," Hoshi said.

Marie pounded her hand on a bulkhead, and Hoshi flinched. "Just give us five hours ..." Marie yelled. "Or I won't willingly board your dammed spaceship."

Hoshi slid off the stool and stood to face Marie. At full height, she was half a head shorter than Marie, but she tilted her chin upward so that her eyes peered down at a woman she clearly deemed inferior. "Malcom, make sure Branson gets safely on board."

"What?" Marie said, suddenly terrified for her son. Hoshi's words stung like a hit to the gut. Marie almost lost her balance, steadying herself by placing her hands on two nearby chairs.

"Everything will be okay," Malcom said.

Hysteric, she lunged for Hoshi, intent on landing at least one, if not two, good punches.

But before she reached the bar, Hoshi drew a gun from beneath her scarf, pointed it at Marie, and fired.

"Oh my God," Marie said, stumbling sideways, a dart protruding from her side. She grabbed the dart, and pulled it out. "You're insa ..." she said, and then dropped unconscious into Malcom's waiting arms.

6

A botanical sphere rested on the northeastern ridge, high above the Martian colony's twelve flexi-glass domes. Autumn had arrived in this tiny sanctuary, where blue birds chirped and squirrels danced on fallen leaves. The breeze carried a hint of petrichor, the scent of cold rain on dry soil, nourishing late-summer flowers.

Scattered along the side of the path were patches of daisies and black-eyed Susan. I knelt down, picking several stalks, and mixed them with wild grasses to form a modest bouquet, tying it together with a ribbon.

A few minutes later, I stood alone at a cenotaph made from red monolithic rock. The monument was engraved with the names of all those who'd lost their lives on this barren planet. We'd added thirty-seven names last month, men and women who died during the battle to save the colony from eccentric-trillionaire-turned-murderer Henry Allen the Third, better known as H3. Henry Allen's hatred for freelivers was reminiscent of the paranoia of history's most horrible dictators. The very thought of H3 made me flush with hatred for the man who had tried to exterminate Martian colonists who had ceased to be "useful."

Ironically, over one hundred of the names were those of Multinational Defense Force soldiers who fought for H3 against the colonists. But in war it was our belief that *all* of the deceased should be remembered.

My eyes traced the lines of a poem carved into a brass plaque embedded into the western side of the monument,

Free from gravity's earthly hold,
And chased by inner stellar imaginings,
We danced through the cosmos on Wernher's fiery train,
A desolate abyss on the shores of a vast eternity,

To greet the magnificent God of War,
A strange red ocean, a rising tide of angry dust.
We are the soon forgotten history,
Sons and daughters of a magnificent adventure.

Though we perished, our story endures of
Boundless frontiers reached not by rocket or booster,
But by childlike imagination, and a reminder that someday,
We'll go home.

~ *Abigail Huang, first human to set foot on Mars - 2035*

What brought me to this place, this hilltop sphere on a red planet? Four years had passed since the cargo ship *CTS Bradbury* crashed and devastated California, when I lost my wife and son. The disaster shook the world, causing earthquakes, tsunamis, and civil unrest. China even detonated several thermal-nuclear weapons in the Tibetan Neutral Zone, as a warning to the Communist Alliance, a group known to use any excuse to increase its influence. Over half a million people died in the aftermath, but the world soon returned to normal. At the time, I was convinced I'd spend my life a drunken freeliver in a Las Vegas hotel.

Instead, I boarded a shuttle, blasting spaceward to the Martian Transport, *Mayflower*. As we sailed toward Mars, some part of me returned; I felt a sense of adventure, wonder, and fear, emotions I never thought I'd experience again.

Now, I reached to my left breast pocket where my sunglasses hung by an arm. In a fluid movement I'd practiced 10,000 times, I flicked open the aviator's golden frame and set the relic upon my face.

My watch buzzed as I walked back to my SUV. "Where're ya, skippa?" came Leeth's Australian accent from my wrist.

"Lost track of time. I'll meet you at the dock."

Exiting the bio-dome through a vine covered door, I entered a translucent flexi-glass tube. The road turned west, passing between the colony's twelve dome-circumferential, and the Alamo, a private dome for the colony's elite.

The vehicle stopped at a pressurized service cylinder just off the main road. Several luxurious auto-cars sat in designated parking spaces, including one finely detailed Electro-Davidson.

The service elevator was almost as old as the colony, and led to the pressurized ice caverns below. At the bottom, the door rattled open and my face was hit with blue light from an artificial sky. A dozen printed J-24s bobbed in the blue water, bow lines secured to the fiber-plastic dock.

I stepped into a sea of socialites, adorned with the latest in sailing fashion, and was welcomed with several pats on the back. Mars's wealthiest residents believed they owed us, crediting us with saving their lives. To repay their perceived debt, we'd been invited to participate in their favorite sporting event, Martian Tube Sailing, or MTS. Avro, Amelia and I had spent hours practicing in VR. Now, we'd get to test our skills for real.

A tall man, with curly hair tied back in a bun, elbowed his way through the crowd of mingling socialites. It was Leeth, the globetrotting (and solar system trotting) Australian nurse. "Hey, Mate!" he said.

"Leeth," I said, clapping my friend on his shoulder. "Is everyone here?"

He nodded towards a boat docked half way down the pier. Avro Garcia and Amelia Shephard checked the rigging while Kevin Patel carried a red cooler into the cabin.

Amelia and Avro had been on Mars for less than six months, but knew the planet better than most. Only a few people complained when the couple borrowed the Arachnid, the colony's most expensive VTOL aircraft, to spend a weekend at an abandoned research station.

Avro reached out a hand and hoisted me down onto the deck. He wore the green jersey of the Mexican National Football team and his permanent half-smile. I gave Amelia a hug, and took a cold beer from Kevin, setting the drink in a cup holder embedded into the hull. Kevin wore a T-shirt from his Alma Mater, Bangalore University. The animation shirt showed an Indian man wearing a Sherwani, high-fiving an Atlas robot wearing a similar Indian frock.

"Ready to race?" I said.

"I was born ready," Amelia said.

The other sailors climbed into their boats and raised their mains. A horn belched and we released the jib, rocketing our slender J24 into the tubes.

At 700 feet underground, the musty air stung our nostrils. It carried with it an earthy scent that reminded us of home. Dense atmosphere pressed against Martian depths as we wrestled the vessel against a backflow of nitrogen-rich air. Moisture dripped from the ceiling as the pressure squeezed liquid water from porous sedimentary rock.

Waves pelted the gunwales and I hauled the tiller to port. The J24 slid around a baffle and crossed into the irons; churning eddies in our wake. Avro released the jib as we tacked. Amelia ducked and let the boom swing over her head. She hauled in the line, winding it clockwise around a winch.

The young Australian sat at the bow, looping his legs through the bow pulpit and sipping Bundaberg rum from a kangaroo skin flask. He closed his eyes and let artificial sunlight warm his face.

We crested a wave and H_2O spewed onto the deck. I shivered as the liquid evaporated from my skin.

Kevin stumbled up from the cabin holding two protein-dogs slathered with ketchup and mustard. My stomach groaned, distracting me from the helm. "If only you could sail as well as you fly," Kevin said, trying to keep his balance as the boat banked around another bend in the tube. Kevin's jet black hair was disheveled in the wind, in contrast to his usual style: fiber-gelled and parted on the left.

"Hike out, Kevin," I ordered. "You're supposed to be our counterweight."

Kevin rolled his eyes and I guessed he figured he didn't spend three years studying robotics to crew a dingy. I cranked the tiller to the right and Kevin stumbled against the rigging, but still managed to keep from going in the drink. Kevin was the lead programmers for the colony's constructor drones; nothing made him happier than new hardware flown in from Earth. He was also the only person on Mars with a bumper sticker: Blueprints to Bootprints since 2035. He inched his way across the deck to Leeth, handing him one of the dogs.

"Our *ballast* needs to lay off the hotdogs," Amelia quipped to Kevin, shouting over the whistling sheets. With sockless feet in deck shoes, Amelia looked as if she belonged on the water. She wore Ray Ban sunglasses and a blue ball cap to shield her eyes from the holographic sun, while Daisy Dukes and a windbreaker completed her nautical look. Only her pale white skin hinted of a life tens of millions of miles from the nearest ocean.

"Cheers mate," Leeth said as he accepted the hotdog. He'd left the clinic still wearing his blue hospital scrubs, tying them in knots below the knee.

Kevin nodded, giving a comic salute with his food hand, then tucked his feet under the hiking strap and leaned back.

Forced air erupted from a vent in the ceiling and a gust struck from starboard as we curved around the next bend. I adjusted our course to avoid careening into the nearest baffle. Amelia grabbed the winch, cranking it clockwise. The jib clung tightly along the gunwale as the boat accelerated to seven knots.

Moments later, it was time to tack again. "Ready about!" I called. "Hard alee!" Amelia removed the lever from the winch and tossed it to Avro, releasing the jib in the process. The boom passed over our heads and Avro cranked in the line to complete the tack.

A rival J24 crossed behind us on the opposite tack, an unwise move considering the tight quarters.

"Hey, circ-dweller," came an accented voice from the other boat. "You're stealing our wind!" I recognized Michael Curry, the man who owned ninety percent of SpaceNet's advertising.

Money was no advantage in MTS; all the J24's had been printed to spec.

"Tactics," I yelled back. Their mainsail luffed, perturbed by the lack of free wind. They lost speed and fell back.

A third boat tacked up-tube at three o'clock; its rainbow jib fluttered as the crew reeled it in. They were in the lead, but not by much. All of their sailors were sober. Our sailors were … well, let's just say we didn't all have that advantage.

Our boat ripped around the final turn like something tearing cloth, and toward an exit arching over chopping water. With one boat ahead and another trailing off our stern, we sailed under the arch, bursting into the Presidio's reservoir. Spectators cheered from a rocky red shoreline. Opposite them, a holographic San Francisco floated on the horizon, with Alcatraz Island shimmering in the foreground.

The three J24's curved in formation as we approached the narrow inlet that served as the finish line. Our boat slid into the lead, but we made toward land and would soon be cut off by the outside boat. I yanked the tiller to the left, swinging around a red buoy, forcing the other boat to turn to port in our wake. If I timed it right, our boat would hide the buoy from view.

The boat banked to starboard, drenching a hiked-out Kevin in a heaving Martian wave. Our boat crossed the irons and Avro secured his jib line, completing the final tack in line to the finish.

Kevin hauled on the hiking cable and slung his soaking body mid-ship. Leeth laughed as Kevin flicked bits of wet bun in his direction. He turned to me, and hucked what was left of the protein-dog at me, and I ducked. The dog flew over my head and landed in the water with a splash.

The other boat cut along to our left, on a port tack, having lost little momentum, unlike us who'd sacrificed precious knots in our last tack.

Jeff Watson, from environmental engineering, kneeled at the tiller. "Sorry boys, looks like we've got you," he said across the gap between our boats. But they were inside the red buoy and didn't know it.

"Watson, tack!" I yelled.

Watson cupped his ear as if he didn't hear me.

I formed a megaphone with my hands and took a deep breath, "You're supposed to keep *outside* the red buoy," I yelled.

Watson turned to look for the missing buoy. Realizing his mistake; he cranked the tiller to the right, but it was too late. The keel connected with a submerged rock and his J24 lurched to port, the mast dipping almost all the way to the water. Watson and two of his crewmates were thrown from the boat.

Out J24 shot through the inlet. There was a horn blast and the race was over.

Avro reeled in the jib, transferring control to the mainsail in preparation for docking at the marina. He smiled and clapped me on the back. "Johnny, someday our winning streak is going to end."

"Not today," I said, giving him and Amelia high fives.

I eased the boat toward the dock and Amelia jumped off. Avro tossed her a bow line which she tied to the cleat with a hitch. Kevin reached out a hand to hoist Leeth from the gunwale.

"So, Mr. President," Amelia joked as we trudged up a flight of stairs to the yacht club. "What's next on the agenda?" She wasn't actually asking a question, but instead imitating Robert Bowden, the anchor from NewsFlash.

I laughed. "Please don't call me that. It's not that I'm not flattered, it just sounds stupid." I chuckled again.

"You *are* the president of Mars," Kevin joked as the animated man on his shirt wrung his wet Sherwani blazer.

"Just because I was asked to join the Martian council ..."

"You're the president of Mars. Just admit it." Amelia elbowed me in the ribs.

The council's previous leader, Edward Lu, stepped down after the last month's battle. When it came time to vote for a replacement, I was nominated. The habitats needed repair and the council wanted an engineer at the helm. The vote was unanimous, and I was elected to lead the council.

We sat at the bar, our backs to a row of pleasure craft rocking in a gentle breeze. A replay of my interview with Bowden ran on a holovision above the taps. "Not again," I said, but Kevin turned up the volume.

"So, Mr. President," Bowden said to a holovised John Orville. I put a hand to my face to hide my embarrassment. "How *do* you plan to deal with the freeliving population?" It was a good question; one that Red Planet Mining Corp's CEO, H3, had addressed in his own unique way. He'd attempted the extermination of all Mars's non-working middle class, people who on Earth would be called freelivers.

I stared at the screen, counting the grey hairs on my hologram's head. "Don't forget, Bowden," my recording said, "a lot has been accomplished since H3's, ah ..." My hologram paused, trying to find the right words for, 'escape as a fugitive.' Bowden settled on 'departure.' "The Panel Distribution Center, of which I *am* still the director, is installing five hundred megawatts of solar, putting many unemployed miners back to work."

Bowden rejoined, "But when that work is complete, what will these people do? They can't go back to Earth. It's a year until the next launch window."

"They came to Mars to work in the mines, many of them second and third generation miners." My hologram took a sip of water. "We're looking at several options. The R&D team has asked for volunteers to join exploration missions. Most of this planet remains unexplored. We're also going to open a new VR training facility for everyone who wants to learn new ..."

Now I'd seen enough; I pointed at the holovision then pressed my hands together. The HV obeyed my gesture and the screen went blank.

"Leeth, did I hear that you and Maranda are back together?" I said, referring the COTS shuttle agent who'd welcomed us in Mars's orbit.

"Yeah mate, we actually think it's going to stick this time; we even talked about getting engaged."

"Well congrats, Leeth."

My watch buzzed for a text: *Pickup at the spaceport.*

"Excuse me," I said, getting up, and tapping my watch to summon my vehicle. "Apparently, I have someone to pick up from the spaceport."

"We're not in a launch window," Amelia said.

"I know," I said. "Excuse me."

"We'll come with you," Kevin said.

"Thanks for the offer," I said, patting him on the back, "but I'll meet up with you tonight."

I left the club, taking the tram to the surface. My SUV pulled up to the curb as I approached. I still couldn't figure out who sent these texts summoning folks to the spaceport. It also occurred to me that since Mars had a functioning auto-car system, why couldn't they drive themselves from the spaceport? Tradition, I guessed.

Workers eyed me as I entered the terminal. The last time I was in this part of the spaceport, Avro and I drove a stolen defense force Jeep through a window. They were putting the final touches on the structure's repair.

A window at the gate provided a panoramic view, and sitting on the tarmac was the most beautiful spacecraft I'd even seen. The underside sported a black heat shield like the hull of a boat. A metallic cabin covered with oval windows reflected the red Martian landscape. The stern housed an array of xenon engines, the type requiring a nuclear reactor as a power source.

The logo on the spacecraft's side read what was the most boring name for an organization I'd ever heard: "Space Corp".

"Space Corp?" I said. "What the hell is Space Corp?"

I expected to hear footsteps from the jet-way, as I approached the craft, but all was quiet. Lights around the hatch blinked green. Pressures had equalized on both sides, and the hatch opened with my approach like a door in a horror movie.

I peered into the spaceliner through the open hatch, stepped inside, and looked around. The spacecraft was empty. A chill ran down my spine, as I stepped forward over the threshold. Maybe the occupant was asleep?

The interior was decorated in tan suede that covered the walls, floor, and the doors. A panoramic window granted extensive views of the spaceport. I walked toward the bow of the ship. Suede VR crèches were adorned with augmented reality visors and resistance gloves. I rested my hand on the back of the captain's chair.

The Martian landscape suddenly vanished, replaced by the image of a man I remembered from a long time ago.

"Norman?" I said. "Norman Kim!"

"John Orville," he said, forcing a smile. Norman was the flight director during my last day in mission control, the day when the *CTS Bradbury* destroyed California. Norman was pulled, paralyzed, from the rubble of the Watney building.

"You're okay!" I blurted. "How are you?"

"Stem cells work wonders, John," he said.

"But …" I said, "you are …"

"On Earth, yes, I am. Norman is watching with a delay. I am his avatar. Please treat me as if I were him," he said, and I nodded. "This Turing computer is programed to respond as I would, and I kind of like being in two places at once."

My mind raced to interpret what was happening: What was a Turing of an ex- NASA employee doing on a corporate luxury spacecraft? NASA had been defunded after the *CTS Bradbury* impacted California, killing over a million people. What the hell was going on?

"I've been watching the news, John. You've become quite the celebrity. Congratulations, Mr. President."

"I really wish people would stop saying that."

"But you've got to admit, it has a nice ring to it," Norman said.

I snickered. "I'll give you that."

"You're probably wondering why I'm here, why this spacecraft is here," Norman said.

"The thought had crossed my mind. What is Space Corp?"

"There is no Space Corp. NASA has been reactivated," Norman said.

I sat on the captain's crèche, leaning toward the screen. If NASA had been reactivated, either they'd been forgiven for the accident, which was unlikely, or …

"*CTS Bradbury*'s crash was not an accident," said the

Turing. "The mission was sabotaged to prevent the construction of the orbital colony."

I was speechless. The *CTS Bradbury* disaster killed my wife and son. Thousands of people went missing after the ship crashed into California, triggering air-shocks, earthquakes, and tsunamis, as far away as Australia.

"I have something to show you. It's confidential," Norman's avatar said. "Only a few people know this exists."

"Why keep it a secret?" I said.

"You'll see ..." Norman's avatar disappeared, replaced by a video feed from several vantage points.

Walter Barrington stood on the roof of his 290 story office tower. A tower he had financed in his younger years for the sole purpose of impressing his fellow New York City socialites. He looked down at One World Trade Center. The old building still glistened, reflecting various diamond shapes from the taller, duller buildings to its left and right.

He paced along the ridge, a knee-high railing separating him from the kilometer-high drop to the street below. Maintenance drones dodged his feet and a cool midnight wind swept over the ledge, tickling his greying beard. He stepped onto the railing, balancing with outstretched arms. Tears traced reflective patterns down his face as he shuffled his feet closer to the edge. He looked to the horizon, and for the first time in years, appeared truly alive as lines softened in his face and his eyes brightened.

Norman spoke above the recording: "That morning, Walter attended his last board meeting, a meeting that would officially dissolve Red Planet Mining Corp., transferring all remaining assets to World Minerals Incorporated. A conglomerate owned by the same group of investors. The transaction did a lot of things; ensuring that any losses incurred by Red Planet during the last two years would be written off the books, and that his family would be taken care of, no matter what happened to him, or the company."

The video feed showed that Barrington's balance was precarious, but he made no effort to readjust his footing.

"I didn't mean for anyone to get hurt," the old man confessed into his watch. "For heaven's sake, the *CTS Bradbury* was unmanned!"

This was a man who still couldn't grasp that *it was all his fault.*

"The probability of hitting Earth was infinitesimal!" he yelled into the wind.

But it did. And hundreds of thousands of people died.

Walter took a breath as if readying himself to dive into a pool. Despite bouncing off an inclined window on the 280th floor, he was conscious until he hit the ground.

The video feed ended and Norman's image returned.

My breathing began to accelerate, and I put my hands up to my face.

"World Minerals Incorporated orchestrated the conspiracy," Norman said. "The 'debris' that caused the disaster was a drone."

"A drone!" I repeated. "Why haven't we heard about this? Why hasn't this been in the news?"

"Things are complicated, John. World Minerals Inc. was highly integrated with foreign governments, not all of them friendly. The official story, according to the US government, anyway, is still that the disaster was an accident, but that the investigation is ongoing."

"So, if I tell anyone, I'll come off as a nut job. Great."

"Unofficially, NASA has been pardoned of all faults in the *Bradbury* disaster, and has been quietly reactivated. There are not many of us old timers left, but we've been tasked with rebuilding the organization. World Mineral Corp.'s assets are now ours. Well, the space-based ones, anyway."

"Was H3 involved with the *Bradbury*'s destruction?" I

asked, feeling sick to my stomach.

"As far as we can tell, no, not directly, though he would have profited from it indirectly. Walter Barrington's confession leaves many things unexplained. We believe H3 might be able to fill in the gaps. We're forming a special task force to find H3, and bring him back."

"What do you want me to do?"

"Remember that computerized selection process NASA used to screen candidates?"

"Yes, of course."

"We're looking for people with deep space experience, NASA background, and for this particular mission, no close family. The computer put your name at the top of the list. In fact, based on our criteria, you're one of the only people who made the list at all."

"I see," I said. "And how, exactly, am I supposed to help?"

"Finding H3 could prove challenging in and of itself, and once we do find him, we'll need your help to bring him in."

"Bring him in?" I said. "We should kill him!"

"We're NASA, John, not vigilantes. Knowing H3, he's probably well-guarded. It could take military force to get close to him, and we've prepared for that contingency.

"We're asking you to come back to Earth, and we'll go from there."

Earth. I thought. *Oh my God, I'm going home.* My mind filled with images of my wife and son, of waterfalls, forests, and open sky, the freedom to run in one direction for more than a kilometer without hitting a flexi-glass wall. But by the sound of it, there wouldn't be much time for sightseeing, and I'd probably be sent straight back to space.

"Alone?" I asked.

"No, we'd like you to assemble a team of two, maybe three, others from Mars, people accustomed to long duration spaceflight.

People you trust. This mission could take months, or even years ..."

"You're excluding people from Earth," I added.

Norman's Turing paused, almost long enough to confer with the real Norman. Finally, the projection spoke. "As I said, H3's network runs deep in the government and private sector. You are correct; we want to keep the pool of people who know about this mission as small as possible."

"I can name a few people who might be in on this adventure."

Norman smiled. "I thought you might."

"When do we leave?" I asked. "The next launch window isn't until …"

"This spacecraft has nuclear ion propulsion. Launch windows are irrelevant. You'll leave as soon as you're ready."

7

Marie dry heaved into consciousness. Her head throbbed and her stomach wrenched. She tried to get up, but something held her down.

"Hey," came a voice belonging to a woman perhaps twenty years older than Marie. "Are you awake?" The voice had a Greek accent.

"Diana? Is that you? Where are we? What happened?" Marie's throat was dry. She coughed; her throat felt like sandpaper.

"Apparently, they gave you some sort of tranquilizer. They said you didn't want to leave, and tried to attack Hoshi!"

"Well that's mostly true," Marie said. "It's a long story."

"The crew asked me to keep an eye on you." Diana sounded angry.

"Where are we?" Marie asked.

"Seriously? On the spaceship."

"Assholes!" Marie yelled, and then started coughing.

Diana frowned and looked away, as though unsure if Marie was someone with whom she'd want to associate anymore.

Marie tried to press her hand to mouth, but something prevented her fingers from reaching her face. She explored her head's enclosure with gloved hands, realizing she wore some sort of fishbowl helmet. Marie looked at the other woman; a non-reflective dome, like that of a comic book spaceman, covered her head. There were dozens, if not hundreds, of people in this compartment, all reclined in dentist chair-like crèches. Above her, several more hung in a web of scaffolding.

Marie leaned back and coughed again. "Where's Branson?"

"Sam called me when he woke up this morning," Diana said. "We couldn't find you; not even my watch knew where you were! How could you leave your own son?"

"I ..." Marie said, her mind still spinning from being unconscious.

"He's with the other kids, where he belongs," Diana answered. "You're welcome."

"I'm sorry," Marie said. "One of the technicians led me to believe I could search for my husband."

They stared at each other for a moment, and Marie watched as Diana's anger faded.

The crèche in front of her was within arm's reach. Marie felt like a slave traveling to the new world in a crate. Fear grew and subsided in waves, her heart beating itself to exhaustion, like an obese man trying to run.

The spacecraft began to rumble. "Launch sequencer is go for auto sequence start," a male voice said from deep within the vessel.

"Hold on," Diana said. Marie looked to her left. A window revealed little information other than that they were still underground.

"Main engine start," said the ship. "Three, two, one." Marie reached for something to hold onto, but grasped at nothing, so she made fists and held them close against her sides. The rumbling intensified. The window filled with fire, and confused flames danced along the silo's walls in figure eights. Her helmet rattled with earthquake intensity. A reverberation, like a fog-horn, resonated through her body, the noise so intense it tugged on her insides, as if trying to pry the vertebrae from her ribs. Marie scrunched up her face and flexed her abs, as if it would ease the pain.

"Liftoff," said the ship, as Marie's body sank into its crèche like a bowling ball settling into a mattress. They were accelerating. Blinding flames illuminated the silo and Marie watched red hot support beams drop beneath them. In a matter of moments, they were outside where the air swirled grey, like the cyclone in *The Wizard of Oz*. Marie felt tears wet her cheeks. She was crying. Any chance of finding John was slipping away along with Earth.

Breathing became difficult as the acceleration intensified. Marie gasped, swallowing tiny breaths of air. She tried to scream, but lacked the strength to force enough air past her vocal chords. She tried to breathe, to fill her lunges, but an invisible weight pressed on her chest, forcing her to release what little air she had. It reminded her of being in labor.

"Max Q," said the ship, indicating the aerodynamic stress had peeked, and the most dangerous part of the launch was behind them.

A minute later, sunlight poured warmth onto Marie's face. She squinted and turned away. Light rays from a dozen windows traced across dull grey spacesuits; it was like a scene from Cirque du Soleil.

The spaceship jolted as the sky outside turned from grey to black.

"Stage separation," said the ship.

"Take deep breaths while you can," Diana said. "You missed the briefing; you'll want to breathe between stages, like this." She took a breath as if diving into a pool, released it, and took another."

Marie forced herself to do the same, watching as spent rocket boosters and fuel tanks plummeted into cloud covered skies. She breathed, submitting to her fear like a child with monstrous shadows on the bedroom wall. Marie concentrated on the crèche in front of her and waited for the acceleration to return.

The spacecraft's second stage ignited. G forces, even greater than before, returned. When Marie was in labor the contractions were minutes apart. She wasn't sure which was worse, but she'd rather be in labor than this. She closed her eyes, trying to offset the nausea.

Inside the helmet, she twisted her heavy head back toward the window. Earth appeared white, covered in clouds. She could see the horizon, a dull brown hue accented with patches of grey, smoke from 1,000 fires.

The spacecraft jerked. Crèches rattled and bodies lurched. "Stage separation," said the ship.

The G forces subsided and Marie cranked her head to the right, looking at Diana. "How many stages are there?" she said, words coming out as a groan. She took a deep breath.

Diana took a breath. "Four …" another breath, "I think."

The next stage ignited and they were pushed back into their crèches again.

After twenty minutes of sickening acceleration, the spaceship's engines stopped, and everything went quiet.

Marie felt as if she had been pulled from the trunk of a car and thrown off a bridge, experiencing that moment in freefall before you hit the water. But the water never came. She looked down toward her feet and felt upside down. She looked up, and felt as if she was spinning.

Earth fell away slowly, like a balloon dropped from a balcony. Marie wanted to throw up, but there was nothing in her stomach. She let her hands drift up in front of her face, studying them, letting the gravity of the situation sink in.

Her husband, John, had dreamed of going to space, but chose a career that kept him grounded and behind a desk. She should have *insisted* John pursue his dreams. *Now he'll never go into space,* Marie thought. Tears pooled around her eyes, unable to run down her cheeks for lack of gravity. She tried to wipe them away, but her hand clinked against invisible glass.

Diana reached over, putting a hand on Marie's shoulder. Other people in the ship talked amongst themselves, creating a moderate din, sticking out their arms and legs to watch them float.

"Take a sip of water," Diana said. "A button, on your neck."

Marie found the button, and pressed it with two fingers as if checking her pulse. A stream of cool water shot towards her mouth and she let it soak her dry throat.

"There's another switch for air, in case you need to evaporate any water inside the helmet."

Marie tapped the other side and a blast of air swirled inside the enclosure, drying her tears.

"Thanks," Marie said.

There was a pop-hiss and the helmets released their grip. Marie reached for her face and lifted the helmet from her head. It dangled in space, connected via a tether to a port behind her back. Marie reached for her face, fingers surveying her mouth and nose; they felt strange, as if they weren't quite connected to her head. She tried to feel her eyes, but something blocked her hand, and she realized she was wearing glasses of some sort.

Marie pulled them off with both hands, hands that were suddenly covered in a thin black mesh. She second guessed herself; a moment earlier, hadn't she been wearing white gloves?

Looking around, she realized several things had changed, though many things stayed the same. Her helmet floated nearby, just as it had, and she was still surrounded by people. Marie looked at the glasses, recognizing the design as augmented reality eyewear.

Something else was missing; the crèches were gone! The passengers around her were fixed directly to the spacecraft by beams that fit snugly into the small of their back.

She looked left; the window was gone. She looked to her right, to where Diana had been moments earlier. There was a body, dressed in black from head to toe. The figure removed the fishbowl helmet. The figure's head was covered by a mesh hood, like a criminal about to commit a robbery. The figure reached up, removing the mesh. It wasn't Diana, but a man with greying hair.

"Diana?" Marie called, reaching under her chin and pulling off the mesh that covered her own face.

"Over here," Diana replied. Diana was situated above, and slightly to the left of Marie. The older woman twisted in place, curly brown hair swirling in zero gravity.

All around the room, people were removing their helmets. Everyone wore the mesh suits, like scuba divers in wetsuits without the tanks or flippers. Marie still felt as if she were seated and strapped in, but glancing down, she realized her crèche was gone. *The suit itself must be providing the feedback,* she reasoned, being familiar with full body VR suits, but never having tried one herself.

A loud metallic screech emanated from the core of the vessel, as though two ships had collided at sea. Marie's body was yanked backward, like a marionette on a swing. Air whistled between the scaffolding. The crowd ahead of Marie moved up and away, yielding a view of the other people in the compartment, and she watched as they spread out in all directions.

"Inflation in progress," said the ship.

The vessel's skeletal structure expanded in a growing sphere. Blue and beige canvas unfolded, forming walls, and rooms. Mood lighting illuminated the canvas in a multitude of colors. As the interior structure spread out, so did everyone attached to it. The spacecraft's interior suddenly looked like a Hoberman sphere, a child's toy that began as a small ball, but could be pulled into a large one.

"What's going on?" Marie asked, as her distance increased from those to her left and right.

"We're in a Bigelow module," Diana said, from several meters away. "The habitat is inflating."

The inflation continued for several minutes, slowing as the ship completed its transformation from the size of a nuclear submarine, to a marshmallow the size of a city block. With the transformation complete, the whistling ceased.

A woman floated in the distance, arching between the scaffolding. A long black ponytail flowed behind her as she pushed off one beam and grasped another. She wore the same VR scuba suit as everyone else, but without the apparatus in the small of her back, the docking port where the suit would connect with the ship.

The flying lady stopped, floating in place. She pulled in her arms and spun around, inspecting the multitudes of people staring in her direction. It was Hoshi.

Hoshi held a small electronic device, placing it over her ear. "Hello citizens," she said. Her voice echoed as if they were in a large auditorium. "Welcome aboard the *Mount Everest.* I'd like to thank our fine crew for the successful launch."

Hoshi paused for an applause that never came. "Please thank the AR team for providing everyone with a window seat. You're probably wondering why we experienced the launch with augmented reality. This was a historic moment, and everyone deserved a view."

Hoshi pushed off again, darting around the spaceship, twisting around columns, and waving at various folks as if they were longtime friends, or if this were the opening act of a circus where she was the lone acrobat. She stopped in the very center of the module, holding onto a beam with her right hand.

"Our voyage to the Jovian system will take three hundred and forty days. Our vessel, the *Mount Everest,* is the largest in World Mineral Corp.'s fleet. But with twenty-five hundred people on board, it will begin to feel very small, very fast. For this reason, we'll live and work in virtual reality."

"In these suits for a year?" Marie said, getting Diana's attention.

"Don't worry, it'll be okay," Diana whispered across the gap.

People all over the ship began shouting questions. Hoshi put a finger to her ear as if she were receiving a message. She nodded, to no one in particular, and then said. "Activate Cali."

Marie yelled, "When can I see my son?"

Hoshi removed her finger from the earpiece. "Pull on your hood, put your glasses back on, and all your questions will be answered." It was an order, but one directed at Marie.

Marie didn't want to put on her glasses; she wanted to escape, strangle Hoshi, and find Branson. But, since that probably wasn't going to happen, she resolved to learn what she could, like everyone else. The hood floated at the extent of its tether. She grabbed the mesh, which felt metallic in her hand, and pulled it over her head. Imperceptible to her sense of touch, the hood touched every inch of her face, with slits for her mouth, eyes and nose.

Marie put on the glasses, and materialized inside a great hall with a stage large enough for a symphony orchestra. Octagonal plates hung from the ceiling like lily pads in an upside-down pond, and colorful tapestries draped the walls. All around her, inanimate bodies came to life,

The seats were red and soft, and Marie could feel fabric under bare legs. She looked down at her body. She wore comfortable shorts and a white blouse, and blue shoes. Fresh air circulated, and the room had a scent, like old fabric. She was seated several rows up, and four seats in from the center aisle. The seat granted a relatively central view of the stage. Pricey seats, had she been a paying customer.

A projection behind the stage read "WELCOME TO SPACE" in large letters and a video, with the caption "LIVE", displayed a spaceship rocketing away from a distant, cloud covered Earth.

Marie studied the spacecraft on the screen. It wasn't entirely dissimilar to ones she'd seen before. The spacecraft's modules were stacked like rounded toy blocks. Next to each module, facts and figures appeared in boxes. The inflatable habitat or 'marshmallow' was white, accented with blue ridges. At 200 by 150 meters, the marshmallow composed over half the spacecraft's volume. On its side, the name *Victoria* was painted in large blue letters.

The marshmallow rested atop a cargo module containing one hundred and thirty-six million kilograms of food. Beneath the cargo module, a service module housed the oxygen tanks, a water recycling unit, and fuel cells. Flanking the service module were two solar arrays, each a perfect circle extending from the spacecraft like a flattened umbrella. Trailing the service module were the fuel tanks and engines.

The video zoomed out, revealing two other ships in the convoy, coasting in loose formation. Marie determined the video emanated from her spaceship, *Mount Everest.* She studied the hulls of the other ships, one named *Klondike*, and the other, *Melbourne*.

Hoshi approached a podium in the center of the stage, holding up a hand to quiet the audience.

A list entitled "FAQ" appeared on the screen. Marie noted the irony: there was nothing 'Frequent' about this. Hoshi began answering the questions as if they were asked by the passengers.

"When can we see our children?" Hoshi said, reading the first line. A video of the ship's nursery appeared on the screen. The children sat in form fitting crèches, each with five point harnesses. Colorfully dressed women unbuckled the kids and placed them on a multi-textured matted floor. When each child was removed from their seat, the crèche slid up into a compartment in the wall. There were cubbies in the room. One of the women opened them, retrieving a collection of toys and books which she handed out to the kids. Marie saw Branson's Washington Capitals' eagle sitting in one of the cubbies.

Marie studied the screen, looking for Branson, and then realized something. *The floor?* she thought. *Gravity?* The nursery was in one of several trams that circled the interior wall of the habitation module. The trams made their way around the cylinder approximately once per minute, simulating the gravity of their destination, Callisto.

Marie spotted Branson, tugging on a teacher's apron. The teacher reached into the cubby, and handed Branson his eagle. Marie realized she'd been holding her breath, and released the air with a sigh.

After giving folks a moment to study the video, Hoshi said, "Citizens will be allowed to visit their children in the nursery, but I'd warn you, the transition from weightlessness to gravity will be nauseating at first, so please, take your time. Early childhood development programs will run during regular business hours. A viewing area will also be available here in VR."

"If I have friends on the Martian Colony can I contact them?" was the next question.

Hoshi passed the question onto Malcom, who stood up in the corner of the auditorium. Malcom's avatar wore a green collared shirt, as if he'd just returned from a game of golf.

Someone passed him a microphone. "Yes, we'll try contacting the colonies. There are radios accessible from VR, and ones on the ship you can use during your shift. However, we're currently limiting long range radio communication until we're away from the Earth-moon system. We're theorizing that the Communist Alliance is jamming communication from Mars and other space outposts."

There was a commotion, but no direct questions. Malcom passed the microphone back, and took his seat.

A woman yelled from the back of the room, "When will we connect with the other ships?"

"Ah," said Hoshi with a smile. "Now that is a fascinating question." Hoshi paused, soaking in the dead silence from the giant room before speaking. "This theater is one hundred seats across and one hundred seats deep. Do the math."

"Oh my god," Marie said, looking around at the room full of people. She turned to the person on her right. "Were you in the

Chinese Hive?" she said, touching a young man gently on the shoulder.

"No, ma'am," he said, "I was in South Africa,"

Everyone began talking to the people around them, asking where they were from, how they'd escaped, and what ship they were on.

"As you've probably just discovered," Hoshi said, her voice loud enough to quiet the room, "all ten thousand citizens are here, right now, in this room."

There was another commotion, as people began introducing themselves.

Hoshi waited until the din began to settle. "This metaverse follows the Schrödinger rule. For those not familiar with the Schrödinger rule, it was named after physicist Erwin Schrödinger's famous cat in the box, a thought experiment where a cat in a box is both dead and alive until you open the box. Nothing, including people, pop in or out of existence while anyone is watching. The virtual experience will be as real as you make it. The claustrophobic days of space travel's past, are over."

Hoshi stepped down from the stage and ascended the center aisle. "Follow me," she said. "You're going to like what you see."

People began filing into the aisles, many taking their first steps in Virtual Reality. Marie stood, not wanting to hold up her row. They followed Hoshi to the exit where she opened the doors in the back of the room, and sunlight streamed in. The crowd poured outside into a pavilion.

Marie passed the threshold, entering a world so breathtakingly real, that all hostility towards her perception of virtual reality faded in an instant. Birds circled in rising currents, chirping as if it were a spring day. Pigeons rested on concrete horses rising from circular fountains. Spanish architecture surrounded them on all sides, while cobblestone streets wound through storefronts shaded by colorful awnings.

In the distance, wind swept across green fields, creating colorful illusions in the grass. A lake reflected the sky and a distant hill. Birds circled on invisible breezes under a clear blue sky.

Hoshi climbed onto the fountain, grabbing a concrete horse by copper reins. She raised her other hand into the air, waiting until she had everyone's attention. Her voice rang through invisible speakers, and her message was clearly heard.

"Welcome to Cali." She paused. "This virtual world is a near-perfect representation of our final destination, Callisto."

8

I waited for Avro, Amelia, and Kevin in the Panel Distribution Centre. As so often when not fully occupied, my mind filled with longing for Marie and Branson. I closed my eyes, and saw them laughing. Nostalgia's illusion favored pleasant memories; any sadness erased by the sands of time. Every day, I did something in their honor, a laugh or deed dedicated to their memory. Today would be no different.

I was jarred from my memories by the arrival of Avro, Amelia and Kevin. Two memorial stones rested on our machine shop table.

They knew what I'd asked them here to do.

We prepped the suits for a very special EVA. This was the last time we'd ever walk on Martian soil. Avro zipped up Amelia's spacesuit, giving her a thumbs-up before working on his own. Kevin checked the comms.

With green light illuminating our pressure regulators, we stepped into the airlock and sealed the door.

I held the flowers gathered from the gardens near the cenotaph, the bouquet of mostly black-eyed Susan and daises. Marie would have been pleased with the selection. The floral scent lingered inside my helmet. When the door rolled open to the Martian atmosphere, the bouquet froze, as if dipped in liquid nitrogen. There was a strange pleasure in knowing the flowers would hold their beauty for quite some time.

Over a kilometer from the colony, Avro parked our jeep at the crest of a hill. We collected rocks and made two piles, as though covering bodies. I'd seen this in an old western movie once, and it seemed appropriate to do now, for my missing wife and son.

As the sun set over the distant horizon, a colorful array stretched to the zenith: blue, purple, orange and gold. Shadows embraced the distant colony like a mother bear protecting her young.

Kevin knelt, running Martian soil through a gloved hand. Amelia wrapped her arm through mine and rested her helmet against my shoulder. Avro reached down, picking up a stray rock, setting it on the smaller pile.

I'm usually not one for talking at funerals. But my grief had reached its limit, giving my mouth the freedom to say what needed to be said.

I knelt at the foot of Marie's grave, resting a hand on the nearest rock. "We were married in the fall, under a gazebo in Arlington, Virginia. There was a breeze and fallen leaves circled around your dress. You laughed. You loved fall, and the leaves tangled in your veil didn't bother you. I lifted the veil, and kissed you, knowing in that moment, I had just made the best decision of my life."

Moving over to Branson's pyre, I knelt again. "I sold my airplane seven months before you were born. Up until that then, that was the saddest day of my life. When I held you and you cried, I knew it was worth it. I swore to never again place value in an object. We're here for such a short time. God, such a short, short time. If there's one thing we should value above all else, it's the people we love. Life is short, so don't waste it."

I removed a Washington Capitals' ball cap from my satchel, placing it on Branson's pile, and then removed my Velcro Orville badge and placed it on Marie's. I stood back, and read the tombstones for a final time.

Marie Orville
2040-2071
Loving wife and Mother
O'er the rivers of your life,
joy and passion flowed
the most powerful.

Branson Orville
2069-2071
Son of Marie and John
Laughter was your pastime.
We'll laugh together,
When I see you again.

We walked silently back to the jeep as the sun dipped below the horizon, bathing our silhouettes in darkness. Phobos and Deimos glided overhead and a small blue planet, Earth, shone brightly on the ecliptic. I studied the planet for a moment, and suddenly I *wanted* to go home.

Avro and Amelia looked at each other and nodded. They were ready to go home, too. Amelia leaned over and gave me a hug.

The next day we took an auto-car to the spaceport. Leeth and his fiancé, Maranda, met us to say good-bye. We trudged through the terminal, wheeled suitcases trailing behind us as if we were a family headed for an Alaskan holiday.

"Sexy," Avro said, eyeing the corporate spacecraft from

behind tempered glass.

"You know, not only does constant acceleration get you there faster, but the one-ninety-ninth gravity experienced on board has a non-zero effect on the body's vestibular system."

"Kevin, English," Amelia said as she pulled her suitcase up to the gate.

"It decreases nausea," I translated.

Maranda used a keycard on the pressure barrier leading to the jet way.

I gave Leeth a hug and slapped him on the back. "Until we meet again." The others exchanged hugs with Leeth and Maranda.

As I walked down to the waiting rocket, Leeth yelled, "Hey!" and tossed me something.

I caught it. It was his flask, full or rum, and I rapped the kangaroo hide. I nodded thanks, and entered the spaceship.

We settled into leather chairs for takeoff and Avro slipped into a suede AR crèche that served as the cockpit.

Kevin secured the door. "Hatch secure."

Avro thumbed the comm. "MATC, this is spacecraft *Mars Force One* requesting permission to taxi to the runway."

"*Mars Force One*?" I said.

"Yes, Mr. President." Avro lifted his AR glasses to look me in the face.

"Oh great, not you too," I said.

The spacecraft bumped along the tarmac as we approached the runway's threshold.

"Rock and roll," Avro said, and activated the takeoff sequence. Rocket engines ignited, thrusting the sleek spacecraft down the runway. It curved up, assisting in the spacecraft's transition to vertical flight.

I watched through holovision-sized windows as we approached the runway threshold. The engine's muted roar was accented by the hiss of thrusters located under the nose. The colony dwindled below us as we accelerated toward space, red skies darkening to black as we climbed.

Main Engine Cut Off, or MICO, occurred 250 miles above the surface at a velocity of over 6,213 miles per hour. Avro activated the magnetoplasma array and set the autopilot for Earth. Plasma gleamed neon blue as the spacecraft began its slow but constant acceleration towards home.

I unstrapped my restraints and floated out of my chair. Amelia did the same, and went to explore the luxurious spacecraft's amenities.

"These bedrooms are huge!" she observed, peeking into one of the rooms at the rear of the spacecraft. "Avro, check this out, there are holomirrors on all six walls,"

"Whoa," Avro said. "Warm up the nuke before you light my fire."

"I call the room farthest from those to rabbits," I said.

Amelia floated back towards the bow. "What's the first thing you're going to do when you get back to Earth?" she asked, resting a hand on the AR crèche as Avro finished up in the cockpit.

"Eat real *carne asada*," Avro said.

"Real beef? I never could taste the difference," Amelia said. "All I want to do is lie on the beach."

"Robot Olympics," Kevin said.

"I don't think we'll have much time for sightseeing," I said. "Who knows where we'll be headed next? I don't think we'll find H3 on Earth."

There was a buzz, and Avro hit a virtual button with the AR glove.

"What is it?" I said. Avro held up a hand, letting me know he was receiving a message.

"Course correction. Zero point, zero, zero, one degrees.

Stand by." He paused as the autopilot plotted the correction, then yelled, "Shit," and tossed his headset at the forward window, and rose from his crèche.

"What is it?" Amelia asked.

"We're not going to Earth," Avro said.

"What the hell?" I yelled.

"That course correction was only a fraction of a degree, we're still *headed* towards Earth," Kevin said.

"That's true, we're headed *toward* Earth," Avro said. "But we're going to land on the moon."

9

Marie had one priority: to find Branson. She flipped up her AR glasses, and saw nothing but masses of people flailing around in weightlessness; nausea returned and she wanted to puke. She set her glasses back in place, and waited in the virtual courtyard for the nausea to subside, realizing that the Virtual World was now an order of magnitude more pleasant than the reality.

She tapped a virtual watch and said, "Where is Branson Orville?"

"Branson Orville is in the pre-school, located at Piazza della Signoria," said the female voice inside her watch. "Head north, seventy meters." Her watch projected a red arrow on the cobblestone street. She followed it, bumping into a lady with a shirt and hat that read "Grief Counselor."

"Can I help you?" the counselor said in a condescending tone.

Marie ignored the woman involuntarily; frustrated that something was blocking her arrow. After an awkward dance, Marie said, "No, thank you," and sidestepped.

Someone tapped her on the shoulder, and she turned, frustrated at yet another distraction. It was Malcom.

"Marie, I wanted to apologize," he said, but before he could say another word, Marie slugged him in the face. His avatar reeled from the blow. Marie cupped her wrist. The punch had hurt both of them. Marie staggered, shocked by the realism of the punch. Malcom cupped his jaw then said "I …"

Marie left him alone on the street and didn't look back. She followed the European terraces to the pre-school, entering a three-story Renaissance structure with several other eager parents.

She found a door marked "Mount Everest Children Ages: 2 – 4." Inside, six children explored a three by four-meter space. The room looked like any pre-school class, with holograms for walls that could simulate a multitude of educational environments. Kid-friendly tablets and large Crayola styluses littered the floor. A box overflowing with stuffed animals rested in the corner. A model solar system hung from the ceiling. Earth, the third planet from the sun, looked just like the old photographs; its blue oceans shimmered as the little globe rotated in midair. Jupiter orbited over Marie's head, its stormy red dot swirling ominously.

The teacher welcomed Marie, introducing herself as Mrs. Shelly Hanson. She dressed the part, wearing a bright red apron over blue slacks and a yellow blouse. The apron's pockets held a stack of Dr. Seuss books, ready to be drawn like pistols in a duel.

Marie sat down on the floor near her son. At first, Branson didn't notice her. She called to him and he toddled over with outstretched arms. Marie reached out and ruffled his hair, amazed at how real it felt. Branson just stared back; his mother's hand a gentle sea of photons on his temples. He didn't feel a thing.

Branson tried to touch her but his hands grasped at nothingness. He reached up both his hands asking to be picked up.

Marie sighed. "Sorry, Branny," she said, forging a smile and backing up to avoid the awkwardness of his hands passing through her avatar. "I can't pick you up."

Branson began to cry, grasping at Marie with outstretched hands, clawing at the projection until he worked himself into such as frenzy that he emanated a banshee-like screech.

Several other parents entered the room, receiving similar welcomes from their children.

Mrs. Hanson gave Marie a tired look, then said, "I'd rather you limit your visits to reality, where you can actually touch your son." She walked over, grabbing Branson's hand. "How about we read a book?" The banshee scream persisted and several other children covered their ears.

Another parent entered the room, a black woman wearing a white jumpsuit. Marie expected her child to go ballistic as well, but the woman reached down, scooping the small girl into her arms. That woman was in reality.

Turning back to her screaming child, Marie said, "Go ahead, Branson, I'll see you soon."

From her pocket, Mrs. Hanson drew *Fox in Sox*, Branson and John's favorite story book. Marie could never get her tongue around it. Mrs. Hanson handled the rhymes without issue, hypnotizing the children with the rhythm of her voice: "a tweetle beetle noodle poodle bottled paddled muddled duddled fuddled wuddled,"—she must have read the book 1,000 times. Branson continued to sulk, but sat Indian-style in front of the teacher.

Marie's watch dinged and a message appeared on her wrist. "Please report to the Center for Genetic Diversity (CGD)." Another arrow appeared on the floor. Marie became anxious, realizing that she was about to meet her colleagues. Meeting new people usually never bothered her, but this was different. These people would be familiar with her work, and she expected them to have strong opinions on it. Diversity of thought was something she encountered frequently in academia, but when it came to defending her work, the debates often made her uncomfortable.

Marie stood up and turned to leave, planning to see her son in person as soon as she got out of this resistor suit. Awkwardness and screams aside, interaction with kids his age was good for

Branson. He'd seen more children in the Hive and here on the *Mount Everest* than he ever had in San Francisco. She took one last look at her son, and walked out the door.

The trip from the nursery took several minutes. Marie was convinced the programmers spaced things out so that residents would get their exercise, combatting the adverse effects of zero gravity.

There were no cars, or trams, or anything else that moved under its own power. Several people had acquired bicycles from kiosks located conveniently around the simulation. The bicycles were blue with friendly baskets at the front and rear. A teenager ripped off the baskets on his bike, tossing them in a recycolizer where they disappeared from existence.

The arrows took her down a switch-back road like Lombard Street in San Francisco. She passed a dozen four-story apartment buildings and two pavilions. The streets were busy, filled with people following arrows. Narrow cobblestone streets crisscrossed in a meaningless manner, convincing Marie this town was based on Naples.

She climbed two flights of stairs leading to a reflective glass cube overlooking a well-manicured lawn. Statues of world leaders guarded paved walkways while flags of the world fluttered in the light breeze.

Marie paused at the top of the stairs then walked toward the structure. On the door, a sign read "Center for Genetic Diversity". Marie sighed before entering, thinking, *I have a bad feeling about this. I possess too much knowledge.*

She pulled on the handle and heaved the large glass door open.

Marie was the last to arrive. She entered a large room unnoticed while the others made small talk at a conference table. The room was lit with natural light (as natural as light in virtual reality could be) from floor to ceiling windows and skylights. Trees cast checkered shadows on the floor. Outside, birds chirped and squirrels scurried around, gathering acorns.

“Hello,” Marie said to the room.

The three others, who Marie assumed were either geneticists or anthropologists, put on smiles, and walked across to meet their new colleague.

James Bekker introduced himself first, shaking Marie’s virtual hand with both of his. He was tall, and despite the grey in his hair, looked to be in his late thirties. His accent was distinctly South African, his body most likely onboard the *Victoria.*

Lise smiled and Marie could tell the smile was genuine, so she smiled back. Deep brown eyes held the pain of loss, but also something else: hope. She was tall, fit, and wore her hair in long braids. Tribal beads hung in layers around her neck. She had caramel skin and a flattish nose. Her bare arms were covered in scribbles and lines in a language Marie had never seen.

“A Native American proverb,” Lise said, aware of Marie’s curiosity. "It says this: ‘When you were born, you cried and the world rejoiced. Live your life in such a way that when you die the world cries and you rejoice’."

“It’s beautiful,” Marie said.

Charles Thomson held his chin down. His eyes looked up over reading glasses that weren’t there, the quintessential look of someone who’d spent a lifetime in academia, brooding over student’s papers with a red curser. His green eyes spoke of wisdom without words. When he spoke, his accent was distinctly Australian, and Marie assumed that his body was probably onboard the *Melbourne*. The man had light brown hair styled with a perfect part down the left side, and Marie wondered if he had styled it “just so” when his avatar was uploaded.

They sat at a table facing a rotating display where 10,000 little identity cards hovered on branches, leaves on a family tree.

Marie sat, staring at the mass of photos and personal information dancing in front of her, the same personal information she'd accessed from her room in the Hive.

"I was in Australia," Charles said. "It was in the wee hours of the morning when the spacecraft hit. Soon after we heard the news about California, a jump-jet landed right in front of my home. A man ran out to meet me, told me a tsunami was coming. I boarded the plane. We met up with several other planes over the Pacific, hovering, and rescuing people from a cruise ship parked near a reef."

"And was there really a tsunami?" Marie asked.

"When the wave hit the ship, it was gone, like that!" Charles snapped his fingers. "And when I say gone, I don't mean sunk. When the wave hit, the ship literally exploded into pieces. Fortunately, they got most of the people off."

"Did you see the bombs?" Lise asked.

"None in Australia," Charles said. "But I was rescued only minutes after *Bradbury* impacted."

"Lise, you're from the *Klondike*?" Marie said.

Lise nodded. "I was on the Hyperloop, heading from Juneua to Vancouver when I felt the impact. They were redirecting trains right into the Hive. I was freaked out when I learned they'd pegged me. Soon as I got off the train someone radioed, 'we got her.'"

"Our Hive was built in an abandoned diamond mine," James said. "I worked nearby at a safari retreat. You are all familiar with the current conflict in South Africa between the local militants and the Communist Alliance?"

"I thought there was a truce?" Marie said, but then realized it was a silly comment. Conflict between the CA and the rest of the world had been marginal, but the atrocities that occurred behind the communist veil, could never be considered peace.

"There sort of was," James replied. "But around the time of the *Bradbury* impact in California, they just started fighting again. I'm not sure why; something spooked them, I guess. There were militants and Alliance fighters in the area, and we could hear them shooting at each other. They started burning all the buildings at the retreat, the hotels, the animal shelters, everything. We evacuated thousands of people down into the mines, that's where we found the Hive. Then, the Doomsdayers showed up, claiming Johannesburg was hit with a nuke and we needed to stay underground."

"I saw San Francisco crumple," Marie said. "Felt the earthquakes and saw the bombs over China. I swear I'll have nightmares for life."

There was an awkward silence.

They turned to the rotating tree of names rising from the table.

Lise put out her hand to stop the rotation. "Let's see what we've got," Lise said. She formed a box with her hands, pulling them apart to magnify.

"Is this live?" James asked.

Lisa nodded. "Yes, and I'm familiar with the algorithm. The simulation predicts population health several generations into the future."

The simulation chimed and a blue circle flashed, drawing their attention to a point deep within the tree.

"A birth?" Charles guessed. "I figured the launch would send a few women into labor."

"Nope, a pregnancy," Lise said. "The system is adapting now." The tree shuffled, and names moved around, tying together the father and mother and creating a new web of possibilities.

James put out both hands and began to shuffle images. "I'll get the hang of this in a moment." He accessed a menu, changing several parameters.

"We can do without the interruptions," Lise said, entering new constraints for the simulation to follow. "I'm going to limit the changes to pregnancy, birth and death. Let's save relationship changes for later."

"Adjust for known cases of sterility," Charles said, not

interacting with the simulation, but following James and Lise's actions closely.

"Good point," James said, typing a new command. "Adding sterility."

Marie leaned forward, witnessing a level of collaboration she hadn't experienced since her work as a post-doc.

"Move post-menopausal women to a different part of the tree," Marie said, stepping up to the simulation and swiping names across the display. "We can still use their genetic material, but only by less traditional means."

The others nodded their agreement.

"How do we know if they're post-menopausal?" James said. "A survey?"

"We can extract that data from the genome maps," Marie said.

"Ah, right," James said. "They're updating those regularly now that we're collecting everyone's urine for water reclamation."

"God, I'd prefer not to think about that," Lise said.

"Can we simplify this? Include only young couples or something," James said, stepping back from the simulation and letting the algorithm run its course.

"We're at minimum population size," Marie said. "We've got to keep all options open."

"Which opens the door to a slew of ethical dilemmas," Lise said. Marie nodded her agreement.

"Like a child with sibling parents," Marie said.

"In a population this size, it's bound to happen," Charles added.

"That child would need to be taken out of the gene pool," Marie said. "How are we supposed to deal with that?"

"That's our job," James replied. "We need to determine what the guidelines are. Only we can answer the ethical questions, creating rules for the continuation of the species."

"What we do here and now will sculpt the society for centuries, heck, for eternity," Charles preached. A profound silence followed.

"What's the goal here? We don't even have any direction!" Marie said. "We'd need to seek input from the populous for nearly every decision we make!"

"No, we won't," James said. "This new society we're creating will operate, for the most part, as a democracy. That said, the Doomsdayers will insure the autonomy of this team will be expressly protected by any constitution."

"How do you know all this?" Marie said.

"There's another reason I'm on this team," James said. "I'm not here only because I'm a geneticist."

"Oh my God," Lise said. "You're a Doomsdayer, aren't you?"

"No," James replied, and then paused to look everyone in the eye. "But my father was."

10

Earth hung in the luxury craft's window like a blue and white marble shining in a dark cave. A hurricane formed in the Atlantic, and green foliage, showing through broken clouds, implied that a rainforest was healthy and well. It was nighttime in the western United States; city lights ran down the coast, only broken by the wasteland that was California.

We'd been in space for six weeks. Despite the luxuries, it still would have seemed cramped if it weren't for the state-of-the-art VR systems. We spent most of our days in a sailing simulation called *Tall Ships*, a role-playing game set in the year 1778, during the privateering days on the North Atlantic. That kept us entertained for the first five weeks anyway; we spent the last week simply watching Earth get bigger in the window.

Orbital insertion began a week ago, three million miles from our destination. Ion thrusters eased us into lunar orbit without much fanfare. In lunar orbit, we received landing coordinates.

The ship rotated 180 degrees as powerful hydrazine rockets brought us down from orbital velocities. The view from the cabin faced opposite the direction of travel. Craters and mountains whipped past, falling over the horizon like ships at the end of the world.

A descent booster roared and the ship settled onto a regolith tarmac illuminated by the faint glow of earthshine. Restraints retracted and cooling thrusters creaked. I got up, standing on shaky legs, popped another anti-nausea pill, and scanned the horizon. A tower loomed in the distance, like a dark silo on a barren prairie.

A rover zigzagged down a path, headlights emanating from its roof. Dust pin-wheeled from titanium tires, dropping to the surface in a perfect parabolic arc. The rover came across our bow; on its hull, the words "moon bus" had been poorly stenciled in blue.

The rover disappeared beneath the transport, coming to a halt below an observation port. A docking ring on the rover's roof kissed the larger vessel's belly with a rubber gasket. We stood around the hatch with our luggage piled beside us.

The pressure equalized and the hatch hissed open. Looking down into the floor revealed a blue ball cap and broad shoulders. I reached down, lending a hand to whoever was below. An arm grabbed my wrist, and I heaved a six-foot man into the cabin.

"Thank you," he said, standing with tall confidence, piercing us from cool blue eyes, the epitome of a leader. "I'm Commander Chris Tayler." A NASA patch glowed proudly on his breast. "Welcome to the moon."

"John Orville." I shook his hand. "This is Avro Garcia, Kevin Patel, and Amelia Shepherd." Tayler shook hands with each in turn.

"After you," the commander said, gesturing toward the hatch.

Inside the rover, two columns of retro bench seats stretched back twelve rows.

"So, Chris, what do you do here, on the moon?" Kevin asked, dispensing with formalities as the commander sealed the hatch.

"Inventory control," Commander Tayler replied, "basically."

"So, you're an accountant?" Kevin said.

"Well, to be fair," Amelia said, "space accountant."

Commander Tayler laughed. "I think you'll be impressed by our new assets."

"What assets?" Avro asked.

The commander gave Avro a self-assured look and said, "You'll see. Hold on." He backed the rover from under the ship, turned, and hit the accelerator. The wheels spun, showering the tarmac with dust, as the rover lurched forward.

We passed the silo, which, on closer inspection, appeared empty. A row of dark windows encircled the structure about half way up.

"That *was* the control tower," Tayler said, "the only surface structure in this facility."

The rover curved around depressions and past towering boulders before descending into crater of a kilometer and a half in width. Its walls were peppered with caverns like a wasp's nest. The rover rumbled along a sloping path turning 180 degrees, and was swallowed by one of the caves.

Triangular walls glowed blue, illuminated by xenon lights. "Access channels," Tayler said. "These lead to the lava tubes."

"Lava tubes?" Kevin said. "As in, the inside of a volcano?"

"The moon's not geologically active," I said.

The rover entered an airlock and a sprocket-door rolled shut behind us. Now under standard atmospheric pressure, Tayler guided the rover along a widening road bordered by rough and unfinished walls.

"When they constructed this facility, they poured regolith cement through the natural channels," the commander explained. "It was a cheap way to make underground roads."

"Who built this place?" I said.

"Care to take a guess?" said the commander. "It's someone you know."

"Henry Allen the Third," Amelia said in disgust. I shook my

head, thinking of H3's attempted extermination of the Martian colonists. The commander nodded.

We rumbled onward for another few hundred meters, and Tayler piloted the rover between two parallel white lines. A claw descended from the anvil-colored wall and clasped the rover's nose. The word "charging" appeared in green on a display.

"This is it," the commander said, throwing a few switches. The rover's interior lights flicked on as the rover's whining actuators dwindled to silence.

Tayler walked toward a dark hallway carved from the rock. We followed, dragging our gear behind us like business people at a sales meeting, and stopped above a machine room with a rail separating us from a ten-meter drop.

"Printers," Avro said, inspecting the machinery below.

Kevin said, "There's nothing special about them; we had a few of these on Mars."

"It's not about the printers," I said. "It's what they're printing." I looked at Tayler. He nodded, with crossed arms, analyzing us as we pieced together the story.

"If this is H3's base. He must have built it to construct his reactor, circumventing the United Nations' ban on launching fission rectors in space," I speculated.

"But you don't make a facility this big for just one reactor," Amelia said.

"Then, what the hell are they powering?" Avro asked, then looked me in the eye, "What do you think, boss?"

"I'm getting chills just thinking about it."

"Hey," Kevin said, "I've been to H3's underground lair. Nothing, I mean nothing, will surprise me."

"I have a feeling we're about to find out?" I said, turning to the commander.

He nodded. "Follow me."

We continued down the hall to a second observation deck. Unlike the view from the previous deck, this rendered us speechless. It was a hangar, large enough to hold a zeppelin. The floor was level, but the sides were rocky and black. Rows of spacecraft lined the walls, arranged like planes on the deck of the carrier. The spacecraft even looked like fighter aircraft, with stubby wings for holding armaments, and tricycle landing gears. The cockpits, or what we assumed were the cockpits, were opaque spheres bisecting each fuselage like a bubble resting on a pond.

"Holy," Kevin said.

"Shit," Amelia said.

"Does each ship have its own nuclear reactor?" I asked.

The commander wore a proud look, as if he'd just became a father. "Affirmative."

A dozen engineers in white jumpsuits scurried around while soldiers in uniform stood guard.

"And those are?" Kevin said, gesturing to the soldiers.

"Marines," Avro and Amelia said in tandem.

"I'm just going to venture a guess here," I said. "These spaceships are nuclear powered, with electrostatic thrusters capable of constant acceleration, high specific impulse well into the five-figure range."

Tayler nodded.

"Small, agile, designed for one pilot," Avro said.

The commanded nodded again.

"Fully armed, and fully operational," Amelia said.

"Nice *Star Wars* reference," Kevin said, winking at Amelia.

I leaned forward over the rail, as if the distance would help with our assessment. "They even have multipole electrostatic radiation shields," I said.

Tayler nodded. "Which means?"

"It means they're designed for interplanetary travel," I said.

Tayler nodded again, but added a smile this time.

"With the latest in onboard VR entertainment," Kevin said.

"Jupiter Jump Ships, or JJs," the commander said. "You are right on all counts. The ships are designed to keep one person alive for up to a year. The pilot inside lives operating within a virtual reality metaverse."

"But why? What are they for?" I asked.

"We're not entirely sure," the commander said. "The ships are fast and well-armed, but aren't designed for operation in atmosphere. We're guessing they're part of a defense system."

"Defense of what?" I asked. "An asteroid? A moon? Or—"

"Or a planet," Kevin finished my sentence.

"Think about it," Amelia said. "What would have happened if H3 successfully conquered Mars?"

"You mean after he's killed the 'undesirable' freelivers?" I said.

"Yeah," Amelia said. "H3 wanted Mars for himself. This was part of his plan. Wipe out the colonists and fortify the planet."

"Possibly," Commander Tayler said. "We considered that. But now we think H3 may have bigger plans."

"What could be bigger than conquering an entire planet?" Kevin asked.

"Come with me," Commander Tayler said. "We have a briefing to attend, and I have some friends I'd like you to meet."

We followed Tayler into a conference room sporting a metallic table and a dozen chairs. Holovisions covered the walls and displayed live views of the desolate lunar surface.

Three other people in blue NASA flight suits stood near the

far wall, two men and a woman, each with aviator sunglasses dangling from zipper-down pockets. They pretended not to notice us. The three talked amongst themselves, pouring coffee and eating donuts from plastic plates. The woman turned, and caught me staring at the yellow wings on her chest. *Astronauts*, I thought, and then averted my eyes.

Tayler clapped his hands onto the back of a chair. “Take a seat,” he ordered. The astronauts grabbed the nearest chairs and stat down. Kevin walked to the refreshments, picking up the donut tray. All eyes were on him as he brought it back to his seat. He set two chocolate glazed donuts on a plastic plate, and slid the tray out of reach. The female astronaut leaned back in her chair, and began clicking a pen, which was interesting, because I was pretty sure the one thing this base lacked was paper.

We sat across the table from our new friends and I read the name patches affixed to their chests: “Singer” and “Nash” and, when I was sure she wasn’t watching, I read “Johnson” off the woman.

Singer wore a skeptical look, like he was about to ask a question, but wasn’t sure what question to ask. When he sat down, he had flipped open a holographic chess game from his wrist, made six or seven moves in rapid succession, against the computer, then swiped the board back into his forearm. He looked around as if it hadn’t happened; he obviously had nerdy tendencies. He was a burly man who looked like he’d just returned from the gym, and with brown hair parted like Captain America.

Nash, a young, bald African American, wore a confident smile, like a test pilot who’d just landed a prototype. He was tall and lanky, but looked like he could outmaneuver Singer in a fight.

Serene Johnson put down her clicking pen, and began scrolling through a report on her watch. She had dirty blond hair tied back in a bun so tight that no stray strands escaped the sphere. The watch’s holoscreen flicked back to Johnson’s wrist and she clasped both hands on the table.

The commander cleared his throat. “Major Singer,” Tayler said. “Introductions please.”

“Major Luke Singer. Former Arctic Command, NORAD.”

“Major Singer is also an aerospace engineer,” Commander Tayler added.

“Colonel Jamaal Nash. Former section lead, U.S. Marine Corps Forces Special Operations Command.”

“Lieutenant Colonel Serene Johnson. Tactical weapons operations. The Pentagon.”

“In case anyone has forgotten, I am Commander Chris Tayler, commander of the Advanced Attack Wing in Charleston, South Carolina.”

Avro leaned over to me and whispered, “I’m sensing a theme here.”

“We are NASA’s latest recruits, the first in the post-*Bradbury* world. In typical NASA tradition, I believe they’ve hired the best, of the best, of the best.”

“The best of the best at what? Killing people?” Amelia said.

“For this mission?” Commander Tayler said. “Yes.”

This was followed by an awkward silence, only broken by the return of Johnson’s clicking pen.

“I’m John Orville, NASA engineer.”

“And, current president of Mars,” Amelia added.

Serene Johnson sniffed a laugh. Nash and Tayler chuckled. They all knew who I was.

“Amelia Shephard, dishonorably discharged member of the MDF.” She seemed to be taking distinct pleasure in her stint as a rebel.

More chuckles from everyone but Serene, who suddenly wore a look of disgust. Perhaps there was something about another soldier disobeying orders that crossed a line for her.

“Amelia is a tactical weapons expert,” I explained. “She was key to helping us save the Martian colony. Before joining the MDF, she was in the army."

Avro leaned back in his chair, looking bad-ass as usual. "Lieutenant Avro Garcia," he said, and winked at Amelia.

"Avro is a SAR pilot, the best there ever was," I said. "He's trained to fly everything from the Martian Arachnid to a 797."

Kevin held up his index finger while he swallowed the last of his donut and chased it down with orange juice. "Kevin Patel. Drone guy."

I caught Serene slowing mouthing the letters O and then K. Kevin always brought a certain awkwardness to first impressions. I'd since learned this was intentional.

Commander Tayler rose and pushed in his chair. "As you may have guessed, we're here because we have intel regarding H3's whereabouts." He tapped the table and the holovision behind him flickered. The live view of the moon vanished and the brown marble of Jupiter appeared, circled by its four brightest moons. "Can anyone name this object?" He gestured at a point of light on the far side of the display.

"Callisto," I said. "Jupiter's fourth largest moon."

"Correct," Tayler said. "Tell us about it."

"Ah, sure, okay," I said. "Twenty years ago, NASA looked at Callisto as a candidate for a deep space outpost. That was before the funding was cut. It wasn't until World Minerals Corp. began sending drone ships that anyone considered going there at all."

"What became of their ships?" asked Nash.

"We don't know," I answered.

"What do you mean, we don't know?" Serene Johnson said.

"The corporations keep their operations confidential in case they discover any large reserves. Since you still can't 'own' space, it's difficult to stake a claim. It could be years before a corporation sets up a colony."

"Thank you, John," Tayler said. He drew his hands together and pulled them apart. The images of Callisto grew, filling the room in front of us. The moon's surface was dark brown, and speckled with thousands of white and yellow meteor impacts.

"These images are from the Tyson Space Telescope. In the visible spectrum, we see nothing out of the ordinary, but here's the infrared."

Tayler hit a button in the table, and Callisto morphed into night-vision green and red. The image centered on the equator, where a fuzzy red band appeared like a line of longitude on a map. The blur indicated the limits of the telescope's resolution.

"Here is the same image over time." The commander hit a button, and a date appeared on the screen: 2068. In this image there was no band. The image progressed to 2069 and a hint of red appeared on the moon's surface. By 2070 the band began to stretch west and by 2071 the band stretched a quarter of the way around. The animation ended in 2074, with the band stretching westward, almost half way around the moon.

"I was at JPL before I left for Mars," Kevin said. "If I remember correctly, we still had a few legacy probes around Jupiter. If we can tap in to any of those, maybe we can get a local feed."

"No," Tayler answered. "The mining corporation terminated all scientific missions when it arrived in the system. Mr. Orville was correct. This is not entirely unusual for these companies, as they didn't want to generate competing interests in the local resources."

"So much for the free market," Singer grunted.

Avro asked, "If that is a strip mine, how is it generating heat?"

"It's not a mine," Tayler said. "We believe they're constructing a *Ring* habitat, three kilometers in diameter. In a few years, the Ring will stretch the entire circumference of Callisto."

"That's crazy," I said. "Callisto has a circumference of, what, over twelve thousand kilometers?"

"Fifteen thousand, two hundred and eighty-eight," Tayler corrected.

"It's R.I.S.C." Kevin said, pronouncing the acronym carefully while marveling at the design.

"Correct, Kevin," Tayler said. "Care to explain?'

"Robotic In Situ Construction," Kevin explained. "This is how the Japanese built the first moon base, with a kind of all-in-one drone. The rocks and other materials pass through a nuclear arc terminal which separates the elements. The slag is ejected and the ore and silicates are used to fabricate the structure."

"Trenching in the front, building in the back," Nash said.

"Exactly," Kevin said. "Theoretically, as long as you provide it with power, the machine can run indefinitely."

"That's exactly what we believe is happening," Tayler said.

"If it's three kilometers wide, what's holding up the roof?" Amelia asked.

"Callisto's gravity is one-ninth of Earth's and it has almost no atmosphere," I said.

"The habitat is essentially one long balloon," Kevin clarified.

"And what happens if it pops?" Amelia asked.

"These things aren't like the domes on Mars, which were just flexiglass. They have spacecraft walls, a series of layers, like the walls on a house. One of the layers is highly compressed vinyl polymer. You could hit that thing with a MOAB and in less than a second the gap would be closed."

"What's it like inside? Do we think it's like the Alamo? Or the Presidio?" I said.

"We don't know," Commander Tayler said. "Several transports left for Jupiter shortly after the *Bradbury* disaster. The flight plan filed by the Jovian Mining Corporation said they were cargo ships."

"Was anyone on board?" I asked.

"Not according to the manifest. But our intel says otherwise," Tayler answered.

"What intel?" Amelia said.

"This. It was picked up a few days before you arrived." Tayler hit 'play' on a recording.

"Lunar station, Lunar station, this is Callisto calling. We are survivors from Earth, if anyone can hear this, please reply. Direct an optical transmission to Callisto, I repeat, the Jovian moon, Callisto."

"Survivors? From Earth?" I said.

"There's more," the commander said.

"This is Callisto station, I may not have a lot of time, a man named H—"

There was an abrupt click, then silence.

"Three," Kevin said.

"That is our assumption," Tayler agreed.

"This sounds like a prank, though," Amelia said. "Why would anyone claim to be a survivor from Earth?"

"It's not a prank," Luke Singer said. "The transmission came via a unidirectional laser pulse. Due to relativistic effects, this would be extremely difficult to fake."

Kevin and I nodded; apparently Major Singer had a background in astrophysics.

"What's the plan, Commander?" Serene Johnson said.

Tayler stood up, and looked around the room. "The eight of us are going to the Jovian system. We're going to find H3. And, we're going find out what the hell is happening on Callisto."

"Oorah!" Nash yelled.

"Oorah," said Tayler.

"On our way to Jupiter, you'll experience the most intense training simulation NASA has ever developed. Johnson, Nash,

Singer, check your egos. You're still ASCANS, just like the noobies. There will be no call signs until you earn them during your training."

Avro cleared his throat.

"Except you, Garcia," Tayler said.

"Thank you, sir," Avro said, smiling in smug approval. Apparently, he'd been called Avro since the day he was conceived.

"Get a good night's sleep folks," the commander said. "We leave at o-seven hundred hours."

11

"Your father is a Doomsdayer?" Lise repeated. "Now I have a HELL of a lot more questions."

"And I'll do my best to answer them," James said.

"Your father is a Doomsdayer," Lise repeated. "And that makes you?"

"*Not* a Doomsdayer. I always thought 'doomsdaying' was just a kooky hobby for the rich. My father made his money in mining."

"You're his successor," Lise said.

"Not exactly, I'm a zoologist, and a geneticist," James replied. "I bring species back from the edge of extinction."

"I bet your father was very proud," Charles said, plastering James with a scolding gaze.

"Here's what I don't get: most rich folks hate freelivers," Lise said. "Yet, half the survivors are freelivers. It's a random sample. Why not be more selective in who they want to save?"

"They selected us," Marie said. "Of all the people in the entire world, they *selected* us, because of our expertise. And there's no other rhyme or reason to the selection of the rest of the population."

"He's just here because of who his father is," Lise said,

looking at the man from South Africa.

"Perhaps you're right." James remained calm, and didn't seem to take offense to the questioning; he probably expected it, had been rolling the questions over in this mind since he entered his Hive. "But my father was different," he said. "His *friends* were different. The Doomsdayers believed everyone deserved an equal chance. They believed a random sample of people was the optimal way to rebuild the human race; not by selecting the people they deemed most 'worthy'."

After an awkward silence, Lise asked, "Where did they get the funding to build the secret colony on Callisto?"

"Space mining has been around since we starting tagging asteroids at the beginning of the century," James explained. "Corporations have been sending stuff into space since before my parents were born, building infrastructure along the way."

"But this ring, this 'Callisto ring' is huge," Lise said. "It must have taken thousands of drones ..."

"Not thousands, only one; A Universal Constructor that we nicknamed the Santa Claus Machine."

"You've lost me," Lise said.

"The machine completes one simple function," James explained. "It constructs a long-pressurized tube."

Lise still looked confused, and was ready to drop the subject, but James continued.

"It's like strip mining. The machine rolls along a track, digging a trench." James ran his hand along the table, pretending to scoop up imaginary dirt. "As the machine travels, it builds the colony."

"Builds it out of what?" Charles asked.

"Elements from the trench," James said. "The Santa Claus Machine refines the mineral, and then mills them into whatever shapes it needs."

"But inside, there are trees," Lise said.

"A thin layer of topsoil," James said. "It doesn't take much organic material to get life started, and Callisto is a very mineral rich world. Last year the Preservation Society sent a small team to get the plumbing going, turn on the artificial sky, that sort of thing. They even placed 3D printing stations along the tube."

Charles's stare dropped to the table, then to the window where the sun hung a few degrees above the horizon. The sky had begun to turn pink with the hues and tones of evening. "I just can't believe they were right," Charles said, "about the end of the world."

"That's where you've got me." James leaned back in his chair. "Doomsday clock aside, I have no idea how they actually knew."

Marie awoke the next morning in her flat, four stories above the central pavilion. Fractured sunlight penetrated the curtains and, for a moment, she wished for shutters so she could sleep a few more minutes.

She threw her legs over the side of the bed, removed the covers from her lap, and tiptoed over the cold floor to the dresser. The clothes in the drawers looked boring and generic; shirts and pants in an assortment of greys and white. She held each item up in turn, tossing it onto the bed.

Now for the part of the morning she dreaded: going to the bathroom. Marie flipped up her glasses and returned to the reality where thousands of bodies floated together in zero gravity. She reached around to her back, and hit the button that released her suit from the scaffolding, freeing her to float to the nearest restroom.

Minutes later, she was back in the comfort of VR. Live video from the nursery displayed on her bedroom mirror. Branson slept alone in the crib-sized cubical where Marie had tucked him in during her evening shift.

She watched as his chest rose and fell.

A message across the top of the mirror read "Two hours per day keeps space anemia at bay". A running stick figure punctuated the sentence where a period should have been.

She riffled through the drawers, finding shorts and a pair of running shoes. Marie slipped on the shorts and put on a shirt. The shoes fit perfectly which surprised her at first, but then she remembered that her life was a glorified video game.

She was about to leave when her watch buzzed, displaying the notification "You have earned $400 eDollars."

"What are eDollars?" she asked.

The watch answered, "eDollars are exchanged for goods and services anywhere in Virtual Callisto."

"Like what?"

"A complete list of goods and services will be available soon," the watch replied.

Marie exited her flat, closing the door behind her. She jogged down the stairs and into the pavilion where several storefronts were opening.

A woman in an apron stood on a ladder hammering on a sign that read "Judy's Art Studio". Inside, rows of blank canvases rested on wooden easels.

Besides Judy's shop, another storefront read "Michelle's Boutique Fashion". A woman inside pulled a blouse over a manikin. The next store over was a bicycle shop; there was a large design screen, and several people stood at the colorful wall, perfecting the designs while a printer brought the bikes into existence.

A quarter mile down the road she passed an open warehouse. Inside, technicians in coveralls danced in front of an interface, using their fingertips to trace the forms the printers would soon create. A tall man in white directed the artistic display, like an orchestra conductor.

"Yes, a little bit more, higher, wider, great … great … now print!" the man said. He glanced at Marie and then went back to waving directions. This was a school of sorts, Marie reasoned, and these people were learning the manufacturing techniques to be used on Callisto.

She jogged along a country road. Pebbles crunched under her shoes, and dust rose in tiny arcs with each pace. Calli, the virtual world which Marie spent over ninety percent of her time in, simulated Callisto's gravity, and Marie flew forward with each stride, feeling like a deer bounding through a meadow. Lush deciduous trees decorated the left side of the road. On her right, green fields rolled to the horizon.

Running past the pillared trees created the illusion that she was traveling faster than she actually was. Even so, Marie was sure that she was achieving her best speeds since college.

She found a trail leading though a forest, and for the first time in weeks, achieved runner's high. She meditated, clearing her mind as the forest blew past. The human mind adapts with surprising efficiency; Calli was beginning to make more and more sense.

Then she thought of John, and how he'd never see Branson grow, or see him learn and experience new things. She wanted to feel the agony of his loss, but the pain didn't hurt as much as it had a few days earlier. She'd also lost her parents, aunts, uncles, best friends, and everyone else she ever knew. Her current indifference to the magnitude of the disaster surprised her, and she thought for a moment that maybe there was more in the water supply than just anti-nausea medication, Valium perhaps. She pounded on, the path sloping upward, with enough incline to make her calves burn.

The road arched over a cantilever bridge and Marie stopped at its center to lean on the railing. She peered down at smooth rocks several dozen meters below. In the distance, the ravine opened into a lake that stretched over the horizon. The sun was still low, but sunbeams illuminated rocky islands speckled with pine trees.

Marie closed her eyes, and let the morning sun warm her body, now fully consumed by the illusion of reality.

She cleared her mind, and counted her breaths, one … two … three …. After ten breaths, tears began flowing from her eyes, fogging her vision. *Tears don't fall in zero gravity*, she remembered. But that didn't matter, she kept her eyes closed, and let the raw emotion course through her entire body. Anger, grief, depression; not directed at anything or anyone, just existence.

Marie thought of John, her friends, and her family, and wished more than anything, that they were there with her. She climbed onto the railing and stretched out her arms, a prayer to the cruel universe that had taken so much. Then, in a symbolic gesture, and without opening her eyes, she dropped into the ravine. She could feel wind ruffling her clothes and for some strange reason, felt at peace. The fall was slow at first, under the reduced gravity, but with each passing second her body accelerated towards the rocks.

Her "death" felt like a belly flop into a pool. When Marie opened her eyes, she was on her bed, face down. She felt pain from head to toe, as if struck by 1,000 rubber bands.

She rolled over, feeling euphoric, got up, threw open the window and laughed at her strange new world.

Marie found herself jogging every morning before waking Branson in his berth.

On this particular day, Marie ran farther than she'd ever gone, and ended up at a shoreline and a sandy beach, where several islands peeked over the horizon under the pink predawn sky.

Marie jogged along a wooden boardwalk. The path curved between sand dunes and over streams trickling into the sea. She noticed a woman standing on the beach with arms raised. The woman began dancing, kicking sand into the air, spinning in circles as if she were a bird sailing in the morning breeze. The movements were fluid, more so than a bird's. The figure twisted, like a fish, arching her body through invisible water in a mysterious dance. The dancing stopped, and she shot her arms behind her back as if spreading an invisible cape.

Marie stepped down from the boardwalk onto the sand. The sun had just broken over the horizon, shining its simulated warmth on her skin. She glanced over at the woman. Long shadows traced her slender figure across the sand, and Marie stopped several paces from the shadow's head, realizing the figure was Lise with her hair down.

She wore a tank top and shorts that exposed her muscular, tan colored legs. Her eyes were closed, and she was taking long deep breaths, letting the light of the new day coat her exposed skin.

After ten breaths, Lise sat Indian style on the sand, keeping her eyes closed; she lifted her hands into the air. After another ten breaths, Lise reached for something tied around her shoulder like a sash. It was a comb on a leather strap. There she sat in the sand, passing the comb through her long dark hair.

She spent several minutes combing the already perfect strands before opening her eyes.

"I sensed you were there," Lise called, with a smile. "I can always tell when I'm being watched."

"I'm sorry," Marie called back, and then began walking across the sand toward her.

"Please, sit," Lise said, bringing her knees to her chest, then stretched them out on the sand, pushing the grains into two damp mounds at the ends of her feet.

"I didn't mean to disturb your meditation," Marie said.

"That's okay," Lise said. "It's important for me to share my people's traditions."

"Oh?" Marie said.

"I am the only Inuit among the survivors. We share our traditions through oral stories, and through song and dance. My father once called me Nuliayuk, the goddess who lives at the bottom of the sea, a place we call Adlivun. Here, Nuliayuk's body and hair flow with the tides. It is in Adlivun that the souls of the people and the animals go to be reincarnated, made into something new."

"That's beautiful," Marie said.

"Look at the sun and close your eyes," Lise instructed.

Marie did and concentrated on the sound of the waves lapping against the shore.

"Feel the warmth of the sun on your face and notice the reflection of the water flickering through your eyelids in soft orange pastel light."

Marie took a breath, exhaling slowly, and observed her heart rate beginning to slow.

"When Nuliayuk's hair is tangled, there is no food to eat, and the sea is violent. To calm Nuliayuk, to calm the sea, to bring the food, I run my fingers through my hair."

Lise pulled her long hair into her hands, ran the strands between outstretched fingers. Marie opened her eyes to observe the process, then closed them, grabbed a lock of her own hair, and did the same.

Marie smiled. "How'd you get here? To this beach, I mean."

"I ran," Lise replied.

"The nearest town is at least eight kilometers away," Marie said. "Where are your shoes?"

"I almost never wear shoes. I like to feel the ground under my feet," Lise said

"I tap out after about twelve miles," Marie said. "Too bad they won't give us superpowers. You know they could."

"They could make us walruses if they wanted to," Lise said.

Marie laughed. "I'm thankful they didn't do that."

Lise got up, and shook the sand from her body. Marie stood as well, reached down and took off her shoes, tossing them away. She nodded at Lise, and turned to run.

"Why did your father call you Nuliayuk?" Marie asked.

Lise laughed. "Because Nuliayuk refused to be married."

12

The quad was scented with the aroma of freshly brewed coffee. I stumbled into the room, tripping into a table with four chairs. Kevin slammed a to-go cup into my waiting hand. I glanced at my watch as my eyes focused. *6:30 a.m.* I was the last to wake.

Avro and Amelia looked tired. My guess was that they had spent the entire night doing … well, you know. Avro turned to face Amelia and embraced her in front of our duffle bag altar. They were lovers, about to be separated, not by extreme distance or comm delay, but by the titanium walls of their spaceships. No amount of VR, no matter how realistic, could replace the spine-tingling sensation of sensual human contact. Or so I thought.

I grabbed my personal effects duffle with my coffee-free hand and held the door open for the others. A woman in a lab coat directed us toward a glass door marked "Preparation chamber". The other NASA crew members were inside, and we nodded our hellos and sat down on parallel benches running along bleach-white walls.

A hidden door opposite the entrance opened and a male technician entered the room. He wore scrubs and protective fabric over his feet. "Right this way," he said. We followed him into a room resembling airport security. "Step into the scanner, please. Leave your bag, and put your hands over your head." The tech was in his early twenties, and nerdy, scrunching up his nose as he talked.

I went first. “You’re not checking for weapons I hope, because I’m clean, I promise,” I said.

“We’re updating your avatar for the metaverse.”

“Oh.” I stepped out of the machine.

“What is the resolution on this thing?” Kevin said as he stepped inside the booth. Sensors encircled his body, the machine beeped, and he stepped out.

“You know that mole on your backside?” the tech asked in a nasally voice. A smile crept across Amelia’s face. “It has two hairs on it.”

“Jesus,” Avro said, shaking his head.

We then entered a room lined with dressing room-like stalls. A sign read “Catheter Station”.

“Oh God,” I said.

I entered the nearest stall and was met by a woman perhaps ten years older than myself wearing scrubs.

“Take off your clothes and lie down,” she directed, extracting a table from the wall. I took off my closes. The woman held out her hand and I passed her each item.

“I’ll get these back, right?” I said as the female technician checked my pant pockets, which were empty. She then rolled my clothes into a ball.

“Nope,” she replied. My face turned red as I handed over my underwear. The technician opened a hatch in the wall marked “Recycolizer” and dunked the clothes inside. The machine hummed as it chewed on the polyester fabric, morphing it into filament for the printers.

I lay down and the woman affixed a catheter and probe to my lower orifices. I grunted as she completed her work.

“Goddammit, man!” Kevin yelled from the adjacent stall.

We gathered outside the stalls wearing only our new high-tech underwear. Amelia and Serene wore hospital gowns. Kevin and

I stood embarrassed, awkwardly trying to look cool by not to hold our hands over half exposed groins. Avro and the commander looked as though nothing had happened, while Singer and Nash wore an oorah-expression as if they'd just returned from battle.

The technicians led us into another room where eight crèches glowed in soft ultra violet light. The crèches inclined at forty-five degrees and inside each one rested a starfish-like entity, resembling a mysterious black sea flower clinging to a coral reef.

Kevin's embarrassment faded and he glowed when he realized what it was. "Military grade feedback suits," he marveled.

"A step-up from anything I've seen before," I said. "Those things look like they can pack a punch, literally."

"That is their purpose," Kevin said. "Literally."

Monitors on each crèche blinked technical statistics. Our names glowed in bold font, superimposed among the data. I walked toward the crèche marked "Orville".

Technicians guided us down into the suits like infants forced into onesie pajamas. The nerdy tech from the scanner, and assigned to me, adjusted the suit around my new techno underwear. My body winced as he placed IVs in my wrists and shivered as he secured biometric sensors to my chest.

Two zippers running up my legs connected at the neck, sealing my body snugly inside. It was a surprisingly comfortable contraption, albeit heavy, even in lunar gravity. After minor adjustments, the tech helped me to my feet. I took a few steps forward with cables dangling behind me like dreadlocks.

We were led to the hangar and lined up against the wall like prisoners facing a firing squad. The techs stood beside us, holding our suit cables off the floor. Commander Tayler stepped forward and faced us. He wore the same heavy suit, and looked like a scuba diver on Trimix, ready to descend deep into a mysterious wreck. Behind him, curious NASA engineers gathered in anticipation.

"Good morning," the commander said, his voice echoing off

distant hangar walls, silencing the growing crowd. "Today we're about to embark on a journey, longer and more dangerous than any mission in NASA's history. We'll face uncertainty and do things no astronaut has ever done before."

Tracking down a fugitive trillionaire? I thought. *Yeah, I'm pretty sure no astronaut has ever done that before.*

"That said," the commander continued, "I can't think of any other people in the solar system I'd rather see in those suits. We've assembled a hell of a team. We've got the best and most advanced equipment. And, with NASA's patches back on our shoulders, I'm confident nothing will stop us. We'll find H3, and bring him to justice." He paused while the engineers clapped. "Good luck, and God speed, ladies and gentlemen. You may proceed to your spacecraft."

The engineers cheered, parting to let us pass as we lumbered toward the Jupiter Jump Ships. Men and women patted me on the back as I walked by, saying, "Good luck," and giving artificial smiles as if we might never come back.

"This is it," my technician said, halting at a JJ. The spaceship loomed overhead. Thermal regulating wings folded upward with rail guns hanging ominously from their roots. Steps descended from the spherical cockpit like the tongue of a cat lapping milk, but pausing to check for danger.

"After you," said the tech.

"Thanks, ah, Hutch," I said, reading his nametag. The name was probably a call sign.

I climbed the four steps leading into the cockpit, feeling the bulk of the suit pressing against my thighs. Inside it was a perfect white sphere, seven feet in diameter with no windows. The rear of the sphere held several cargo compartments, which popped open when touched. I stashed my personal effects bag and turned to lend a hand to my tech.

Hutch was right behind me, holding the cables. He pressed

his hand on the side of the sphere marked "SPAR INTERFACE" and a metallic beam extended into the cockpit from the rear. A digital interface appeared on the wall and the engineer began entering commands.

"Place your back against the spar, please," Hutch said. I set my back against the beam as instructed. He tapped more commands and the floor began to rise, creating a stool beneath my feet.

Hutch pushed on my midsection, forcing my lower back deeper into the spar. My suit click-locked into place with an electromagnetic seal.

"I'm going to lower the stool now," he said.

I nodded as the stool sank into the floor, leaving me hanging in the air. The technician connected the cables, verbalizing each cable's function as he plugged it in. "Power, check … Data, check … Fluid, check … IV, check … Waste management, check." He paused, and then said, "Okay, you're all set. When you arrive on Callisto, you'll find the release switch in the main menu."

I continued to hang, waiting, while the technician retrieved one more thing. The hood. It was black like the resistor suit, and covered my entire face. He pulled it down over my head. A cold metallic mesh contacted my skin, touching my cheeks, forehead, nose, and lips. My eyes alone were left uncovered. With the hood in place, the tech sealed a panoramic visor over my eyes, plunging me into darkness.

He made several final adjustments to my suit and asked, "How's that, everything snug?"

"Yup," I answered. "It's snug alright. But kind of dark in here."

"Hang tight while I activate the system."

There was a pulse of electromagnetism, the sound of capacitors filling with charge. My suit went rigid, my legs pressed into a seated position by some invisible force. The visor flickered on and I could see the hangar. I could see my hands, my arms, but not the suit, or even Hutch. My arms were bare. I could see veins and tiny hairs on the back of my hand. The wedding band wrapped around my finger, permanently etched in the Avatar's micro-polygon structure. It was the one thing I never took off. I thought it fitting that even the computer considered it a part of my body.

My avatar wore a blue NASA jump suit, with black boots laced up in military style. I could feel the warmth of wool socks underneath.

Semitransparent displays flickered in augmented reality: navigation, communication and flight controls. Glass rudder pedals shimmered into existence below my feet. I reached forward with my right hand, grabbing the control stick, which I could feel in my palm.

Technicians descended from the cockpits of the other JJ's.

"I'm going to leave you now," Hutch said. "Good luck."

I could hear him, but not see him, jog down the steps to the hangar deck. The hatch hissed shut behind him. The tech shimmered into existence as he walked further from the spacecraft, into range of the ship's external cameras. The engineers funneled out of the hanger, into the complex, like school children leaving the playground.

A message flashed in my field of view: "Comm System Active". The words were accompanied by two bursts of simulated radio static.

"Commander Tayler here. Comm check, over."

"Johnson here, I read you five by five," Serene said. Her voice emanated from the left side of my cockpit, her spacecraft illuminating as she talked. According to a giant number hovering above her ship in augmented reality, her spacecraft was designated "2". Tayler's spacecraft was number "1". My spacecraft was designated number "6".

"Singer, five by five." Singer's voice emanated from the number "3" ship.

"Nash, five by five," said number "4".

"Shephard, five by five," Amelia said, sounding confident as usual.

"Orville, five by five," I said.

"Patel, five by five." Kevin sounded scared. I didn't blame him.

"Avro, five by five."

Tayler's ship illuminated. "The NASA folks have control. They'll activate your reactor and load our training programs for the trip. After the ion engines have been activated, they'll transfer control back to us. Any questions?" A momentary silence lingered over the channel. "Excellent. I'll meet you in the metaverse. Enjoy the launch. Tayler out."

The spacecraft taxied from their berths, turning down the length of the hangar. The launch bay door opened wide, swallowing the first JJ before closing like a giant's mouth. Thirty seconds later, the mouth opened, empty, ready to swallow its next meal.

I watched Amelia's spacecraft disappear behind the giant's metallic jaw. Less than a minute later the door opened and my JJ lumbered inside. The barrier closed behind me and the air drained from the room.

A forward hatch opened and my spacecraft lurched forward. The deck beneath my craft pitched upward, and I faced down a tube, feeling like a spitball in a straw. LEDs blinked inside the tube-like runway lights. Stars twinkled at the end, their perfect light polluted by cryogenic outgassing in the chamber.

"Liftoff in five, four, three, two, one," said my ship's female computer, a voice identical to the Sam Turing computer found in most holovisions.

Hydrogen thrusters fired, throttling up to full. *Here we go again,* I thought, and flexed my stomach muscles in anticipation of the massive g-load. Mag-brakes released and my JJ shot forward like a bullet. The craft rumbled and my vision blurred with the vibration. The acceleration pressed me into my virtual seat, the resistors fighting back, preventing me from being skewered by the spar stemming from my back.

Within seconds, I was thousands of meters above the lunar landscape, then kilometers. The mountains and craters blurred behind me.

My heart pounded, as my lungs fought the acceleration. With the sun behind me, the blackness of space filled my peripherals, the VR helmet forming the image from dozens of cameras on the spacecraft's exterior.

A display pegged my relative lunar velocity at a relatively paltry 10,000 kilometers per hour, a speed just north of lunar escape velocity. A moment later, the acceleration stopped and weightlessness caressed my body once more.

"Ion engines activating," said the ship. I felt a slender push as the nuclear reactor came online, the VASMR engines creating an electromagnetic field in the spacecraft's rear. Xenon from the spacecraft's tanks trickled into the engines. Even with nuclear powered ion engines, it would take our ships over a week to reach our cruising speed of 300,000 kilometers per hour.

The eight ships met in formation as we screamed away from the Earth-Moon system. The Jupiter Jump Ships were stealthy, absorbing 99.9% of all light in the visible, infrared, and radio spectrum. With boosters switched off, the ships were effectively invisible against the blackness of space.

"Spacecraft control transferred to internal flight computers," Tayler said. I could tell there was a hint of a smile in his voice. "It's time we settle in and get to know each other. You might want to close your eyes for a moment, it's about to get very bright."

The dark view of outer space disappeared, and suddenly, my

eyes were awash with light. Kevin raised a hand to cover his.

"There are sunglasses in your pockets," Tayler said. The day was warm, but a cool breeze blew against my skin. *The suit uses temperature and pressure differential to create wind,* I reasoned, reaching into my left breast pocket to pull out the aviators and thinking about how silly it must look in reality, a gloved hand reaching at my chest, grabbing at the air.

We took it all in. Palm trees bent towards the shore, providing shade over a half dozen white plastic reclining beach chairs.

We stood on the porch of a bamboo house with a thatched roof. The structure lacked window panes, and we could see inside. The door frame, which was in serious need of paint, lacked a door.

The eight of us stood on the deck in a circle. The men wore Hawaiian T-shirts and trunks, and the women wore short-shorts and floral blouses.

"What is this, 1950?" Serene said.

"There are plenty of other clothes in your closets if you don't like what you're wearing," Tayler said.

We followed him through the front door and stood in the living room. Assorted chairs, that appeared to have been purchased from a thrift shop, faced a coffee table made from two electrical spools, obviously dragged from the sea, their wooden planks turned grey like driftwood. The walls were decorated with surf boards and guitars. Single incandescent light bulbs hung from black wires.

"What about sleeping arrangements?" Amelia asked, looking at Avro.

"Everyone has their own room, but if anyone wants to bunk together, that's fine with me," the commander said, going to the kitchen and pouring himself a drink.

Avro touched the small of Amelia's back. She smiled, pleased with the sensation.

I looked at each of the bedroom doors that surrounded the living area. Our names were painted in pastels on driftwood hanging from frayed rope. Near the back of the common area, a pool table, ping pong table, and a foosball table beckoned to be played.

"Oh, I'm totally killing y'all at foosball," Amelia said, going over to the table and twisting a couple of the knobs. Kevin followed, retrieving a ball from a gumball machine located by the back door.

Luke Singer and Jamaal Nash sat together on the couch. A holovision in the corner turned on and began playing a basketball game, Cleveland vs Miami.

I walked over to the commander. "Is that a beer?" I asked.

The commander smiled, and tossed me a cold one. "Nonalcoholic," he said.

I twisted off the lid and put the bottle to my lips. I could feel liquid emerging from the seam in the hood where the resistors met my lips. The liquid was chilled, and flavored. "How many flavors-sims are there?"

"There are enough," the commander replied, his tone indicating that he'd spent more than his fair share of time in deep-sim.

Serene gave the commander a nod, and he tossed her a Heineken. She took it over to a hammock and sat down, feet on the floor, and stared into space. I wondered what her story was, if she'd lost people in the *Bradbury* disaster, too.

I paced around the room, stopping at a bookshelf resting above a cabinet filled with volleyballs and tennis rackets. The titles were arranged in alphabetical order, starting with Asimov's foundation series, and ending with several titles by Jules Vern and H.G. Wells. I pulled down a collection of Hugo Gernsback's *Amazing Stories*, and opened the anthology at the center. The images were of cartoon shaped rockets, and women in metal bathing suits brandishing oversized ray guns. I sat down with the book, turning the pages between sips.

By the time I finished the beer, I'd almost completely forgotten I was zooming through space, alone in a seven-foot sphere.

When everyone was settled into one activity or another, the commander held up his hands, pausing the game on the holovision.

"Enjoy it while you can, folks; training starts tomorrow morning at zero-six-hundred. Wear shoes, not sandals, because once we begin, you won't have time to change."

A watch on my left wrist indicated the time; I looked at it, and did a double take. We left the quad at 0600, were in space by no later than 1100.

That was less than an hour ago. I looked out the window. The sun hung low in the sky, and it was almost evening.

We were now on ship's time now, but 0600 would come far too soon.

13

Another five months passed and the convoy reached the halfway point in its journey. Marie and Lise jogged through the Park of Nations near the Center for Genetic Diversity. Flags of the world fluttered in the breeze and birds soared overhead.

A bird-cycle race passed over. The bicycle hang-glider hybrids granted far greater maneuverability than the hang-gliders. Their pilots peddled like madmen, as a simple gear drove the canvas corkscrew, a system inspired by Leonardo daVinci; a machine that would only fly in *this* gravity.

Lise was usually quiet on their runs, using the exercise as a form of meditation. But this morning, she said, "I met someone."

They jogged an additional five paces before Marie responded. She was caught off-guard by the sudden change in character. Lise didn't seem like the type of woman to get excited about a man. "Let me guess, you were testing that compatibility application you designed?" Marie said, slightly out of breath from the run.

"Okay, I admit, it was a selfish endeavor," Lise said with a smile, and not a hint of exertion.

"Does he know he's part of an experiment?"

Lise made a face, as if shrugging, but with her eyes, and

Marie laughed.

"There's something else," Lise said. "He's on my ship, the *Klondike*. We've been sharing meals together, and spent time outside of Calli in the *Klondike*'s private rec room."

Marie stopped running and Lise hauled to a stop a few steps ahead, turning around.

"You …" Marie said, leaning over to rest her hands on her knees, a natural reaction to stopping after a long run.

"Won't be skinny for much longer."

"Pregnant," Marie said, and paused to process a strange emotion. For the first time it almost felt, not wrong, but odd, to bring a child into this new world, a world without Earth. They'd convinced several women to have children once they arrived on Callisto, filling gaps in their genetic algorithms, but Lise would be the first one she knew personally.

"I guess you're doing your part for the algorithm, too," Marie said.

After a hug and a laugh, Marie started jogging again, and Lise followed. They passed a lake surrounded by weeping willows. A man stood in waist deep water, fly fishing under the willowy canopy.

"You should try it," Lise said. "The compatibility application, I mean, at least see what it comes up with."

"Not going to happen," Marie said.

"Would *you* consider have another child?" Lise asked. They crested a hill above the lake, and slowed to a walk to catch their breath.

"Our job is to preserve the human species. I'll do my part just like everyone else." Marie and Lise both knew what the algorithm recommended; anyone able to have children should, and there wasn't much room for debate. Marie had been thinking about this for quite some time.

The thought chilled her to the core. If John was here they'd do their part, no questions asked. If one of them died young, would the other remarry? *Maybe.* But when a person goes missing, in a way, they are neither dead nor alive. Marie thought of the Schrodinger rule that applied here in virtual reality and how it seemed to apply to this situation as well.

Lise touched Marie on the back of the arm. "Promise me you'll let me know when you're ready to pick a donor."

"It'll be years before I'm ready to consider it. But I promise, I'll let you know." Marie looked into Lise's eyes. "Eventually, we all have to do our part."

Several people on bicycles zoomed by, calling "on your left" as they passed.

The birdcycles arced overhead and circled around a soccer field where two teams battled over a multicolored ball.

"Look! There he is!" Lise said, pointing out at the field.

"The one on the green team with the yellow socks?" Marie said.

"Yeah, that's him!" Lise said, her face glowing.

"A fine specimen," Marie said.

"Want to meet him?"

They went over to the stands and poured paper cups full of water from a cooler. Marie took two swigs, her resistance suit squirting liquid into her mouth.

Lise shouted and waved her arms, cheering for the players.

"What's his name?" Marie asked.

"Brian," she answered. "Excuse me." Lise took a half breath and began to cough. "I think I need some more water, my mouth … so dry…"

She stumbled back over to the water cooler and leaned on it. The cooler fell over, spilling its contents onto the grass. Marie ran over and placed a hand on Lise's back.

"Are you feeling okay?" Marie asked, guessing it was

morning sickness.

Lise's hands shot up to her neck. She turned and looked Marie in the eyes. Lise's eyes bulged as if she had just been punched in the gut. Her mouth opened and closed like a fish's out of water. It looked as if she was trying to scream but nothing came out.

"Lise!" Marie yelled, then turned. "Help help! Something's wrong!"

Marie looked over to the field, and saw Brian running toward her. He stumbled to the ground, hands shooting up to his neck. His eyes bulged, too. His hand shot out toward Lise, and then he lost consciousness, lying on the ground, eyes unblinking.

Marie ducked as a birdcycle crashed into the ground, missing her by only a few feet. Its occupant ejected from the seat, slamming into the ground and sliding to a stop. Marie looked up as several others hang gliders and cycles fell from the sky, one of the machines impacting with the soccer net.

Marie held Lise in her arms, then laid her on her back and began doing compressions. *Could a resistance suit initiate CPR?* she asked herself.

"One, two, three, four, five," Marie counted compressions as she'd been taught. "Help," she cried. She looked around as she continued her compressions. Several other people were doing the same. Several had left VR, leaving their zombie avatars to look for someplace to disappear from existence.

People shook the bodies and told them to wake up. On the soccer field, a quarter of the players lay on the ground, unmoving.

A momentary flash of light and the sky turned off. Marie looked around, noticing that it wasn't just the sky that had vanished; it was everything. She stood there in the blackness, her suit no longer providing any feedback, and she was suddenly very aware that she was weightless.

She reached to flip up her glasses, but stopped as a computerized voice spoke in her headset. "Resetting Program."

Three seconds later, the light returned, and Marie stood in the bedroom of her apartment. She left her flat, bolting down the stairs to the street where several others stood around looking lost.

"What happened?" Marie yelled to a man across the street. He just shrugged.

"Marie," someone yelled. Marie looked around. "Marie!" the voice called again. It was Diana. The two women found each other in the crowd.

"Look, that's Kathy, from the *Victoria*. There's Mika, from the *Melbourne*. But Phillip and Kai, my neighbors from the *Klondike*, they didn't appear. Something's very wrong, Marie."

"Is anyone here from the *Klondike*?" Marie yelled. Diana and Marie hurried through the crowd, asking each person which ship they were from.

Marie started running down the street. "Is there anyone from the *Klondike* here? Anyone?" she yelled.

Marie got to the end of her road, where the town ended, and the park began. The distant hills pixelated, the lake ceased reflecting the sky, and structures began to vanish. Marie screamed, her body jolting as the suit demagnetized, leaving her weightless once more. Panic ballooned in her chest. *Branson!*

Again, she reached for her glasses, to return to the reality of the spaceship, but once more her suit went rigid, this time pressing her into a seated position. The light returned and they were in the auditorium. She looked around and saw thousands of people seated as if they were about to watch a symphony. Something was different than before; a quarter of the seats were empty. Several people began to scream at each other, and at the center of the room.

On the screen behind the stage, a camera followed Hoshi.

She climbed up the stairs that led to the stage, walked to the middle, and stood silently as if collecting her thoughts. She then took three steps forward, and sat on the edge of the stage, letting her legs dangle over. Her image projected up onto the screen, and the room went silent.

"As many of you know, there is an asteroid field located between Mars and Jupiter," Hoshi began. "We're passing through that field now. The probability of a ship impacting an asteroid is minuscule, even for a ship of this size."

There was an eerie silence in the room.

"We saw the rock on the radar thirty minutes ago, but there was a fault in our asteroid deflection cannon, and it didn't fire. We started the boot up procedure on the thrusters, and they came online, but it was too late. The asteroid was the size of a baseball. It penetrated the *Klondike*'s hull, passing through several of the compartments, including the O2 supply and hydrogen tanks."

Hoshi paused, before stating in a slow and methodical tone, "All lives were lost."

14

Commander Tayler set our alarm clocks to 0530. I stumbled out of my room, reaching for a mug and pouring myself a coffee. An oval tube snaked from behind my head, squirting warm, flavored liquid into my mouth, and I wondered if it actually contained caffeine. I set the mug down and looked at Amelia. She had dark patches around her eyes and I remembered that there were cameras in our headsets and we were looking at each other's real eyes.

At 0600 we gathered on the beach. Serene wore the same short-shorts and I subconsciously focused on her toned legs. They reminded me of Marie. My wife had run five miles a day up until Branson was born. When I glanced up at her face, she noticed me checking her out and rolled her eyes, as if to say, "Don't even think about it." I would have told her not to worry about it but lacked the energy.

"Walk with me," Tayler said. We followed him along the sand. Kevin, Avro, Amelia and I to the commander's left, and Serene, Johnson, Jamaal Nash, and Luke Singer to his right. A flock of pelicans passed above us, wings beating the air with a ferocity we could feel. The tide clawed at the shore, reaching up the beach and drenching our shoes.

"Our mission is full of unknowns. But if the MDF presence on Mars was any indication of what these trillionaires are capable of,

Callisto's inhabitants might blow us out of space the moment we arrive. I don't want anyone thinking we're wasting our time."

Tayler stopped and turned to face us. "We could teach you how to use these Jupiter Jump Ships to their full potential. And we will, in time. But today, and tomorrow, and the day after that, we're going to be learning something far more important."

He was silent for a moment, and I wondered how many times the commander had trained for battle. "We'll learn to work together as a team. We'll learn to fight together as a single cohesive unit. We'll even learn to die together."

"What happens if you die in the simulator?" I asked.

"We can die in the simulator?" Amelia said.

"He means die in simulation, like dying in a video game," Avro said.

Kevin raised a hand. "You have a question, Dr. Patel?" the commander said.

"Where do we go when we die?" Kevin said.

"Patel, you'll have your answer within the hour. I guarantee it."

Kevin gulped and Commander Tayler smiled. We could tell he was up to something.

"Look at your watches and tell me what you see," the commander said.

"It's six forty-two," Serene said.

"And what else?" Tayler asked.

"It's December seventh?" Luke Singer said, glancing up from his watch. "Which is strange, because that's not the date."

"Hey, John isn't that your birthday?" Kevin said.

"Uh, oh," Avro said. "What year is it?"

"Ah, 2074?" Serene answered. "Why?"

"No, in this VR program, what year is it *supposed* to be?" Avro clarified.

Tayler grinned. "You'll find out in about thirty seconds."

We looked at each other, half expecting a sea monster to rise from the surf. Serene squatted like a tennis player waiting for the serve. Luke Singer and Jamaal Lawson stood back to back, ready to fight. Twenty seconds passed and nothing happened. Then sand began to vibrate as if a backhoe just plowed over nearby dunes. The distant hills began to grumble like giants digesting their lunch. A flock of birds in a perfect V disappeared and reappeared beyond a speckling of cumulus clouds. Then another V appeared, and another.

Avro answered his question. "Well shit, Commander. I'd say this is 1941."

I looked at the airplanes. "And this is Honolulu."

The first Zero emerged from behind the tree line; its approach had been hidden by shoreline palms. The silver plane released a bomb. The shell cracked through the thatched roof of our beach home and blasted through the living room floor. I clenched my teeth in anticipation. The bomb was time delayed.

The house exploded in a ball of red and yellow flames. Bamboo shot through the air like spears from Troy, the blast pelting our faces with debris and heat. Flaming two-by fours cartwheeled into the sand and ocean.

Two Zeros dove toward us from the west with guns spewing fire. Bullets puckered the sand in two sets of parallel streams that arced across the beach. We scattered toward the palms like hockey players racing across the line.

"This way," Tayler ordered, and began running along the hard dirt between the palms and the sand. A Zero approached from the sea, spitting fire from its guns. I skidded to a halt as black volcanic earth exploded in my path. The others paused too, letting the bullets pass.

The commander led us down a path through the trees until we burst into a clearing, an airfield. The low repetitive pulse of anti-aircraft fire hammered on my eardrums as soldiers ran across the field, firing rifles at the sky.

A fighter plane idled on the runway. The Wildcat's yellow-tipped propeller spun slow enough that the blades were almost visible. A white and blue star on the fuselage reflected the morning light; this plane was the pride of the American Navy. The pilot throttled up, and the plane rolled forward, accelerating down the concrete runway. A Zero banked overhead, nosing down, and riddled the Wildcat with bullets. The American aircraft veered off the runway, bounced across the field, crashed into a maintenance shack and burst into flames.

A soldier slapped a fresh ammunition canister in to a 40mm anti-aircraft gun and pounded the sky. The attacking Zero fell to pieces overhead, its wing breaking apart at the root. It spun into the ground, collapsing in a mangled pile. Only the tail, with red circle and stripes, survived the crash.

Double speakers hanging from the barrack's eaves sounded the message, "Air raid Pearl Harbor. This is not drill."

"Get to the hangars," Tayler yelled. "You'll find aircraft fueled, armed, and ready to go."

"I don't know how to fly!" Kevin yelled.

"See those two B-17s?" Tayler said. "Patel, take the one of the left and I'll join you. Shephard, you take the one on the right, you just became tail gunners. Johnson, join Amelia, you just became a pilot."

"All due respect sir," Serene said, "I know how to fly a damned aircraft."

"Aren't we a little short staffed?" Luke Singer said, pointing at the squadron of attacking aircraft on the horizon.

Tayler ignored him and said, "P-forties are in the hangar to the east. Wildcats to the north." He pointed at the hangars as two of them exploded in a mushroom cloud of flames. "Scratch that. No Wildcats."

Amelia and Kevin ran toward the B-17s as the rest of us ran for the P-40s. Technicians were everywhere, loading ammo, and running for cover.

I climbed the ladder of the nearest Warhawk, pulled on a leather headset and tested the comms. They were pre-set. I hit the starter as the Allison-engine's twin superchargers caressed my ears with the most glorious sound a pilot's mind can fathom.

The P-40 had six fifty-inch Browning machine guns. I felt the trigger under my finger.

Tayler's voice crackled over the radio. "Orville, Avro, flank the torpedo bombers from the ocean."

"Copy. Flank the bombers," Avro said, as we taxied across the tarmac.

Commander Tayler's voice came again, "Nash, Singer, cover the Fortresses."

I taxied by Serene's aircraft and watched as she settled into her cockpit. Her eyes narrowed in concentration as she flipped multiple switches, spooling up the four cyclone rotary engines. Something about a woman behind the controls of a powerful aircraft was very seductive. But maybe I was just on an adrenaline high.

I taxied left and right, trying to see around my 1,000 horse power engine. "Shit Avro! There are holes in the runway," I yelled.

"Fast taxi, single file," Avro said, "Stay on my six ... [radio static]. There's the clearing … [radio static]. Take off!"

Avro and I punched in our throttles as Singer and Nash took to the runway behind us. The anti-aircraft gun did its job and no Zeroes bothered us as we took flight. I pulled back on the stick, raising the nose forty-five degrees. The island sank beneath us as Avro and I climbed at full power.

Below us, Singer and Nash circled the airfield while Serene Johnson and Commander Tayler lumbered into the air in the B-17.

"This is a mess, over," I said watching squadrons of enemy aircraft make their way toward the harbor.

"Stay high and to the east," Avro said. "We'll hide in the sun. Make strafing passes, try not to sacrifice altitude for speed. After each pass, head back into the sun."

"Copy that," I said, lining up for our first pass. Avro went first, lowering his nose and hitting the throttle. I followed at his four o'clock and pressed the trigger. Flames shot from the three barrels on each wing. The plane recoiled. I applied forward pressure on the controls to compensate. Japanese gunners took aim from rear canopies and fired back. Bullets arced through the air, flickering against a background of sky and earth.

Smoke billowed from enemy bombers like steam from a locomotive. We pulled out of range and took inventory of our targets. Three began falling in an accelerative dive, while a fourth entered a spiral, having taken a hit on the starboard aileron.

We lined up again for another run. "Avro, we've got fighters approaching from the North. Fight or flight, buddy? Over," I said.

"Finish this run then follow me."

Avro lined up for another attack as the fighters approached at high speed. He strafed six of the bombers, sending a few well-placed rounds into each plane, but not enough to guarantee a kill. I hit five in a similar fashion. Smoke poured from the Nakajimas. We'd hit fuel tanks and hydraulic lines, but the targets stayed airborne.

"Stay on my six," Avro said. "We're going in."

"Going in where?"

He twisted his P-40 to the right then left, descending to the bomber's flight level. I followed a few plane lengths behind. He soared between the larger aircraft, opening fire whenever a bomber crossed his path. The Zeros orbited above the bombers, holding their fire to avoid shooting their comrades.

The first wave of bombers released their ordinance over the harbor.

"Follow the bombs," Avro yelled. "There's nothing more we can do from here."

We dove, placing the bombers between us and the pursuing fighters. At sea level, the *USS Arizona* took a direct hit. Its hull bucked and the giant battle ship began to list.

North of the harbor, a swarm of enemy fighters narrowed in for the kill. Nash and Singer circuited the slower B-17s in a wide arc. Bullets spewed at random targets from Amelia and Kevin's tail guns. They were in rough shape. Serene and Amelia had lost two of their engines. The commander, with Kevin in tow, banked toward the sea with two Zeros in pursuit.

"Hang on, Amelia, we're coming," Avro said.

"I don't think you can save me this time, baby," Amelia said.

"Oh, would you guys shut up," said another voice. It was Jamaal Nash. "You know we'll be sucking back mojitos in an hour."

"Major Nash, that's incorrect," Commander Tayler said. "You'll replay this battle until you win, over."

"You've got to be kidding me, over," Nash radioed back.

"You're stuck in here kid, Tayler out."

Avro and I dove toward the B-17s while another wave of twenty Zeroes approached from the west.

"Hey guys, you're about to have company," Avro yelled.

The approaching Mitsubishi fighters created a veritable wall of bullets, but Nash didn't see it. He banked into the rain and his aircraft was torn to shreds.

Singer banked into an inverted dive, trying to get away from the onslaught.

"Where the hell are you going, Luke?" Serene yelled.

"It a lost cause," he said.

"Get back in the soup, Singer," the commander ordered.

The approaching fighters opened fire on the Flying Fortresses, tearing off their tails, wingtips, and flaps. The bullets dug giant gaps in the sides of the planes, allowing visibility all the way into the cockpit.

"Mayday, mayday, may ..." came Serene's voice over the comm.

Serene and Amelia's aircraft impacted with a hillside, exploding in an orange ball, followed by a mushroom cloud of black smoke.

We expected a similar call from the commander's aircraft, but when I looked over, his cockpit was gone. Kevin's tail gun continued to blaze as the aircraft jackknifed into the harbor. Pieces of wings and fuselage skipped across the water like rocks across a pond.

Avro and I circumnavigated a beehive of enemy fighters, weaving to avoid crashing into the other planes, and firing our guns dry. I looked up in time to see Avro get shot though the cockpit. Blood splattered over the inside of the canopy, leaving me alone in the simulation.

To my right, a Zero followed me into a tight left bank. He fired, but his bullets whizzed under my fuselage, unable to find a home. I leveled out and pulled up. The pursuing Zero did the same, flying in formation off my port wing. I studied the Japanese pilot, his white bandanna with red sun clearly visible behind polished glass.

"Screw it, if they can kamikaze, so can I," I said, banking my aircraft into his. The Zero's propeller sliced into my wing, tearing it off. My plane began to spin. The other Zeros continued to fire at my P-40, bullets riddling the plane's body as it continued to fall. My canopy took several hits, pelting me with glass. The resistance suit attacked my body with electric shocks and kinetic stimulation, rendering the feeling of being stabbed by multiple knives. I screamed. It hurt, but it could have been much worse; adrenalin masked much of my pain.

With my aircraft in tatters, I unstrapped my harness, brought my feet up onto the seat, and jumped. The plane was spinning and I missed the wing as I entered freefall. The beach approached fast and I pulled the rip cord, releasing a parachute at 200 feet. I fell into shallow water, my feet hitting the ocean floor. I tried to breathe, but couldn't. My feet kicked up a cloud of sand, pushing against the bottom, and shooting me up to the surface. The parachute floated nearby. I tried to swim but the cables held me back until I pulled the release and headed to shore.

The battle was over. The enemy planes had already headed back to their aircraft carriers floating north of the islands.

I walked toward the beach house, my body pulsing from the assault of a suit that apparently wanted me dead. The house still smoldered, but the porch was there. I climbed the steps to the deck.

With a flash of light, I was back inside the JJ, then we were back on the porch. The Beach house returned, just as it was before the battle.

"What took you so long?" Avro said.

"I didn't die," I said.

"You didn't die?" Serene said. "You lucky bastard! We just spent the last hour in purgatory, waiting for you."

"Purgatory?" I said. "Let me guess, you were floating in space, with VR off?"

"Yeah, purgatory," she said.

Commander Tayler gestured us inside for a quick debrief before we did it all over again.

"So, what did it feel like?" I asked as we stepped into the cottage. "Dying, I mean."

"Like drowning," Kevin said, "while being hit by an auto-car."

"Like being squashed," Amelia said. "Claustrophobia, times a million."

15

An old man began printing headstones, on each one writing a short dedication. He planted six rectangular headstones on a hill on the outskirts of the town. Walnut trees shaded the impromptu cemetery, and a light breeze ruffled their leaves, allowing sunlight to trickle to the ground and paint the grass in a kaleidoscope of green and brown hues.

The hill was visible from a road near the town. By noon, others had joined the old man, writing eulogies on fiber-plastic stones, then strolling through the walnut forest to find the perfect resting place. By the end of the day, the hill was covered with graves, and if one were to count, they accounted for almost all of the *Klondike*'s 2,500 passengers.

It was the day after the *Klondike* disaster and Marie spent most of it in her room. She lay in bed, staring at the ceiling, trying to instigate sleep in an attempt to quell the pain. But sleep never came. Depression set in, and Marie mentally added the *Klondike* disaster to a list of tragedies that would haunt her forever. She began to wonder what the point of living was. Why not just finish them off, and be done with it? By the afternoon, she was overwhelmed by hunger. She punched out of VR and hugged Branson until he cried.

She tucked him into bed and read to him until he fell asleep. Marie stared at her son; his chubby cheeks, and flutter of eyelashes.

This is why I am doing this. Branson's life is worth more to me than all of humanity combined.

Marie left the nursery, floating between canvas walls to the cafeteria. About 100 people gathered around the tables, feet tucked into footholds to keep them from floating away. Someone was handing out pens and paper, and it seemed as if everyone was having the same conversation: what to write about their friends.

With a full belly, Marie returned to VR with a new perspective. She walked to the hill and the printing kiosks brought there for the sole purpose of printing tombstones. Marie typed ten words into the machine: "The life in her years was bountiful. Lise Anne Locklear." Marie made sure Lise's headstone had a view of the east, facing the sunrise.

She climbed to the crest of the hill and looked back at the town. The sun rested on the western horizon and buildings cast long shadows across the park. The Center for Genetic Diversity glittered in the evening light, its glass façade reflecting nearby trees. Things would be very different when she returned there tomorrow.

Marie took her seat around the conference room table with Charles and James. At first, no one spoke. Clouds obscured the usually blue sky, and strange shadows drew swatches of darkness across the usually cheery space. Next to Marie, Lise's chair sat empty.

"We lost twenty-five hundred people on the *Klondike,*" Charles said. Marie exhaled a huff of air, annoyed that someone would state the obvious, but realizing that someone had to break the silence. For this team, the loss of 2,500 people was a practical matter, population being the most important metric in the genetic algorithm.

"That puts us well below the optimal count," James added. "We've deleted the *Klondike*'s population from the system. And we've instructed the algorithm to construct a sustainable population using seventy-five hundred people. The issue with the algorithm is that it's Swiss cheese, full of holes."

"True, but one of the reasons we recommend ten thousand people as minimum is because it takes disasters like this into consideration," Marie murmured. She hated how easy this had been as a thought experiment, and how gut-wrenching it was in reality.

"It's not that simple," James said. "The computer is basically frozen, saying 'what now humans?'"

"I think we're giving folks too much freedom here. With our current restrictions, there's no way the algorithm will work. Not without some serious alterations," Charles said. Then he stood up. "We should mandate reproduction from the bank and restrict natural reproduction. Assume total control over the next generation."

"We're not going to force husbands and wives to accept embryos from the bank," Marie said. "Humans choose the DNA for their kids by selecting mates. I refuse to *mandate* anything!"

"Marie, you know this is what we have to do," Charles said. "There's no other way. We've got to plug the holes."

"Forget the holes," Marie said. "The first priority is to get us back to our original population. The holes in the algorithm will remain, but a population increase will buy us time."

"That's an understatement; if we get back to our original population, it will buy us a generation," James said.

"A birth boom?" Charles asked

"Yeah, well, a baby boom," Marie said.

"How do we force people to have kids?" James wondered.

"We just do!" Charles sounded determined.

"No, we don't. At no point do we sacrifice our values to save society."

"He's right, Charles," Marie said. "If we don't have values, what do we have?"

"We'll have less people like James," Charles joked.

"Not the time," James said. "Marie, I'd like to hear more about your ideas."

"I don't believe we need to force people to have kids," Marie said. "I think we can get everyone to do it willingly. People just need a nudge. The founders of this mission haven't used force to get us to do anything. Well, besides forcing me to come along."

"They didn't force us!" Charles said.

"They forced me," Marie said. "Hoshi shot me with a tranquilizer in the Hive when I tried to stay."

James chuckled at this. "I know Hoshi, she's fanatical all right. I've heard her say 'The ends justify the means' once or twice to my father." James paused. "She needs you, you know. Badly."

Marie rolled her eyes. "Listen, we've got a few months left until we land on Callisto, and we need every woman who can get pregnant to be pregnant. I recommend we *name* the upcoming generation. Give it a historical significance."

"You mean, call it Generation Hope or something like that?" James said.

"Exactly like that," Marie said and forced a smile at James. She could tell from his tone that he liked the idea.

"So, what's next? How do we gain support for this Generation Hope?" James said.

"With the first baby," Marie replied. "Well, pregnancy anyway."

Charles and James stared at Marie as if they were teenagers and had just found out they'd impregnated their girlfriends.

"After I'm pregnant, we'll announce: Generation Hope."

"You're going to have a baby?" James said, his eyes wide, like a proud father's.

"Yes, James, I'm going to have a baby."

Marie's watch chirped and she rose from her bed in a strange delirium. She usually awoke long before her alarm and went for her run, but feelings of anticipation and dread had kept her awake until deep into the night. Marie hated it but knew that it was necessary. She hoped this baby would also give her hope for humanity, the way that Branson did. She skipped breakfast, and headed to the ship's medical bay located in the core of the *Mount Everest*.

The core wasn't much more than a canvas cylinder punctuated by openings, as if a worm had munched on it like an apple. But it was subdivided into various facilities. She glided by a lady pulling along a bag of laundry, containing the white gowns worn by folks who spent any amount of time outside VR. As the lady passed, the smell of recycled polyester wafted in her wake. She smiled, pointed Marie to the clinic, and went on her way.

Inside, a nurse instructed Marie to wait in an empty holding area painted in a calming yellow hue; images of smiling babies adorned all six walls. Canvas chairs stuck out from two of the walls, their teeter-totter like seats allowing the user to sit without floating away.

A video played on a holovision with images of children in a sandbox, narrated by a corporate male voice: "Humanity has the technology to allow parents to choose the color of their children's eyes, hair, and even skin. However, due to resource constraints, we are currently unable to accommodate customizations."

Another video played with the same narrator: "Fifty years of research has confirmed that zero gravity gestation is safe. Requests to relocate to the centrifuge nurseries will not be granted. If your baby is born on the spacecraft, you will move to a designated nursing station." Marie wondered if the videos were customized for her.

A female nurse in white scrubs floated to the door, her long brown hair dancing as she came to a stop. "Marie, Doctor O'Brian will see you now," she said.

Marie nodded, released herself from the chair, pushed off the wall, and floated behind the nurse to the examination room.

"Put this on and wait here, please." The nurse handed Marie a gown. "The doctor will be with you shortly." The nurse floated away, pulling a blind across the room's entrance.

An examination table extended from one of the walls. A refrigeration unit glowed in blue light. Marie wondered if this was where they stored genetic material. She slipped off her jumpsuit, and wondered if the Doomsdayers had specifically targeted doctors as they had targeted her. She donned the gown, storing her single piece of clothing in a cubby on the wall, and then sat on the examination table. She pulled an elastic around her waist, and slid her feet behind a bar.

Doctor O'Brian arrived without a smile, closing the blind behind her, and stuck her feet to a Velcro pad on the floor. Marie wondered if the doctor spent much time in Calli, and if she knew anyone from the *Klondike*. The doctor was an older lady of perhaps seventy. Her hair was short, curly blond, and bounced in zero G. She held a tablet with Marie's file, and flipped through the files with a swipe of a finger.

"That was fast. Usually I'm waiting forever to see a doctor," Marie said in a quivering voice.

"You're from the Genetics Team," Doctor O'Brian said, reading from the file. She looked up. "I'm sorry about the loss of your colleague. Lise did fantastic work."

"She was a close friend," Marie said. "I promised her she'd help select my donor."

"In a way, she will," the doctor said. "The algorithms your team designed will choose a donor on your behalf."

"Right," Marie said.

The doctor went back to swiping, pausing to read, and then swiping again. "The system has a match for you. Do you want to know anything about the father? Race, hair color, anything?"

"No," Marie said abruptly, and then faked a smile.

"Are you sure you're ready for this?" Doctor O'Brian asked, putting an awkward hand on Marie's knee. Marie stared at the hand until the doctor removed it.

"It's something I have to do," Marie said. "We need to rebuild the population, and I need to set an example."

"A child is not an example," O'Brian said.

"John, my husband, and I talked about having another child. I've been ready for this for quite some time." Marie was lying about being ready, but she didn't know if she would ever be, not without John. She knew after carrying a child for nine months, the child would be hers, and she'd love the child like she loved Branson.

"Would you like to know the sex?"

"No," Marie said. "I'll wait until birth. It was a tradition in my family."

Marie readjusted herself on the examination table, wrapping her arms under the provided supports and lying back. "I'm ready if you are, doctor."

"You're feeling well, no nausea?" James said.

"No, ah, not yet, it's only been a day, James," Marie answered. "Are you okay? You remind me of my husband when we found out. You seem excited, or nervous, I'm not sure."

"Maybe a little of both," James said. Marie was seated at a table with her back to the window. The flags in the park were visible through the curtain of willow tree branches.

"It's time to spread the news," Charles said. "Bringing your speech up now." He typed a few commands into the table and the words hovered in large black font on the wall opposite Marie.

"Where's the camera?" Marie asked.

James laughed. "We're in VR; there's no need for a camera, but if you like we can project your interview on a holovision so you can see what you look like."

"Sure," Marie said.

"Just let me know when you're ready, and we'll record your statement."

James sat diagonally across from Marie. He leaned forward, giving her an encouraging smile.

Marie nodded to Charles, then faced the wall and took a breath.

"Hello citizens. I'm Doctor Marie Orville, a generic anthropologist from the University of California, Berkeley. I'm here at the Center for Genetic Diversity with my colleagues, Mr. James Kotze and Doctor Charles Thompson.

"I want to thank all of you for helping us these past months by providing information regarding the plans for your family. Your cooperation helps ensure that we'll have a heathy population for generations to come.

"But like many of you, we lost a friend and coworker. Doctor Lise Locklear was on the *Klondike*, and now, our work has gotten much more difficult. Ensuring genetic diversity with a smaller population isn't easy and we need your help." Marie swallowed a lump in her throat. "I lost my husband on Doomsday." She was amazed how saying it felt like being punched in the gut. "I lost my best friend on the *Klondike*. We have a plan that won't replace those we've lost, but it's a plan to honor them. We call it Generation Hope.

"Earlier this week, I made the decision to have a child in honor of those we lost. I went to the clinic, and they well, they did their thing." Marie paused to give a sincere smile. "Last night, I took the test, and it's confirmed, I'm pregnant.

"If he's a boy, I'll name him John, for my husband, and if she's a girl, I'll name her Lise, for my best friend."

Marie looked at James and Charles, a tear dripping down her face. Though they were smiling, she could tell this was emotional for them, too.

"So, here's my request of you. Please talk to your families, spouses, significant others. Talk to your ship's doctor and talk to your friends. Let's rebuild humanity together.

"Do your part in creating Generation Hope."

The feed ended and Marie got up from her chair, walked over to a window, and cried.

The next day, Marie received a text from Doctor O'Brian. "You're going to want to see this," her watch said.

Marie tore off her headset and flew to the core. The nurse met her at the door with a look of optimism that had been rare among the survivors. "Marie, we've had a line out the door all day. You did it. It looks like Generation Hope will be a huge success."

16

After losing the battle of Pearl Harbor four times in twelve hours, Commander Tayler granted us five hours of sleep. I rolled around the cabin's cot, wishing he'd let us out of VR so we could rest in the blackness of our sphere. My body ached with the abuse of dying. The electrical and temperature stimulation was far worse than the kinetic impacts, and I was sure there'd be scarring. The first morning light trickled in and I moaned as I realized it was time to do it all over again.

We stumbled onto the beach like slugs. Avro looked as if he were still asleep. Tayler began another one of his speeches, one of those speeches that was supposed get us all pumped up. It didn't work. He talked, but I didn't even hear the words.

I sucked back the last two inches of coffee in a brown mug, and chucked it into the surf. The mug would materialize again after we were dead. Kevin, Nash, and Singer hucked their mugs as well, letting them smash against a wet rock poking out of the water twenty feet from shore.

My heart began to pump faster, as the hum of approaching Nakajima bombers grew.

"Are you sure these resistor suits can't kill us?" Luke Singer asked, rubbing his ribs.

"They won't cause any permanent damage," Tayler said.

"Tell that to my spleen," Kevin said. "I swear, I'm peeing blood."

"We're outnumbered, fifty to one," Avro said. "How many times do we need to do this?"

"I really don't see how we can beat these odds," I said.

"We have an advantage," Tayler said. "We've experienced the battle before. We're beginning to anticipate their every move. I've seen each of you act on instinct. During the last run we took out a fifth of the Japanese air force."

"Yeah, then I got kamikazed by two planes at once," Kevin said, giving himself a fist pump explosion.

Tayler ignored Kevin's remark. "What we're learning goes beyond instinct. You're learning to be creative and you're learning to function as a cohesive team. Creativity and teamwork, ladies and gentlemen, wins battles when you are outnumbered fifty to one."

The commander was right on one point, we were getting better. The enemy attacked in a similar fashion each time. They still reacted to our every move, but became more predictable.

"Avro, you're leading this round," Tayler said.

Avro nodded and turned to face us. "Anyone have any ideas before those Zeros blitz the beach?"

"What if we radioed the ships," Amelia said. "We have them shoot up flack in advance of the attack."

"KP, can you handle that, boss?"

Kevin scowled, arms crossed. "Sure, boss, they'll definitely listen to me," he said, in his strongest South Asian accent.

"We could jerry rig the bombs to explode at high altitude," Luke Singer said. "We'll fly a B-17 to thirty thousand and drop timed fuses over the incoming squadrons."

"Can you rig the fuses?" Avro asked. Singer nodded.

Kevin uncrossed his arms, revealing a t-shirt that showed a race between a trio of magneto-cycles.

"Kevin," I whispered. "Where did you get that shirt? There wasn't anything like that in my closet."

"Ever hear of SpaceNet?" Kevin said.

"You found a way to access the internet from purgatory?" I said.

Kevin winked.

The moan of the first approaching Zero returned to destroy the beach house and we instinctually turned to run for the airfield.

Kevin grabbed my arm, holding me back. "Remember the Schrodinger rule," he said as the pieces of the house rained down around us. It wasn't a question.

I shrugged. "What the hell, Kevin?" I said.

"This is virtual reality. If you can't see it, it doesn't exist."

The next two Zeroes came in for their strafing pass and I turned to run. It sucked to get shot and I wanted to delay the pain. Kevin stood on the beach with his hands in the air, waving at the Zeroes.

"What the heck, Kevin!" I yelled.

"Schrödinger!" Kevin cried in my direction as bullets puckered the sand at his feet. Four rounds ripped into Kevin's chest and his virtual body exploded in a shower of blood and guts.

"Idiot!" I yelled.

"Shut up and run," Amelia said. "The sooner we're in the air, the sooner we can strategize."

We sprinted to the airfield, and everything appeared as it was the day before. *What was Kevin talking about*? I thought, but then thought of something else. "Take cover!" I yelled. "Incoming!"

The six of us dove into a machine gun bunker; when I was sure everyone was in with their heads down, I yelled, "Clear!"

We jumped from the bunker to continue running toward the aircraft. Whatever Kevin was planning, had worked.

Suddenly, the technicians weren't running back and forth

like before. Instead they stood holding their tools and ammo boxes, looking confused. One of the Turings simply dropped their ammo box and stared in awe.

On the tarmac rested row upon row of F-35 fighter jets and B3 bombers. The fighters were gray, and sat ready with canopies open and additional air to air missiles hanging from their wings. Their tails were proudly decorated with the blue and white crest of the United States Air Force Materiel Command, 412th squadron. There were more modern fighter aircraft, but this was the last generation that actually required a pilot.

Stairs led up to the B3 bomber's modern cockpits, cockpits flanked by twin hypersonic intakes for the four wave-rider engines. There was no need for a tail gunner on the B3; the aircraft could effectively outrun bullets.

"Ah, guys," I said, "I don't remember the Americans having *jets* in World War Two."

"They didn't," Amelia said. "Frickin' Kevin."

"Same plan as yesterday," Avro said, ignoring the fact that we were obviously cheating. "John, you're my wingman. Amelia, you're with Tayler." He tapped Serene on the shoulder. "Johnson, you okay piloting a bomber by yourself?"

She nodded. "I wish you assholes would stop asking me that."

"Nash, Singer, cover them."

Avro and I sprinted up the ladder into the F-35s. The cockpits were simple, with wrap around touch screens containing almost everything the pilot needed. I placed the augmented reality helmet over my head, briefly acknowledging the irony that I was using AR in VR. I hit a switch marked "canopy." The hydraulic dome came down over my head and digital information appeared on the visor, as well as a camera feed from around the aircraft, allowing me to see through the plane. The plane came to life with the depression of a red button. The single Pratt and Whitney engine roared to life, and I eased the throttle forward, taxiing away from the other aircraft on the tarmac.

"Forget the runway," Avro said in the radio. "See that lever above the throttle? That's the hover control."

I pushed the lever forward. Hatches on the bottom and top of the F-35 opened to reveal the plane's primary lift fan, while the tail nozzle tilted downward.

Easing the throttle brought the F-35 gracefully into the sky. Avro was more confident. He slammed his throttle forward, shooting skyward and reaching 500 feet ASL in seconds.

"Follow me!" he yelled, taking his plane out of hover, and engaging the afterburners.

Oahu Island sank below me as shadows from the approaching squadrons drew flickering patterns on the hillsides. I pulled back on the hover control and the aircraft shot forward.

The first squadron of Nakajima Torpedo bombers banked toward battleship row. They'd never make it.

"Keep up, John," Avro said. "You don't want to miss the show." He soared toward the squadron of Nakajima torpedo bombers and highlighted the leading aircraft using a touch gesture on his forward display. The internal weapons' bays opened to reveal six air to air missiles. He released all six in sequence. The weapons streaked toward the advancing squadrons, plowing inside the attacking bombers and detonating inside the core.

The fourth generation AIM-9 Sidewinder missiles were designed for much larger enemies, and the Nakajimas disintegrated as our weapons erupted. The resulting explosion could level a neighborhood. Surviving bombers disappeared into giant balls of flame. Most made it through without exploding, but all were mangled, and some were on fire.

Pilots and bombardiers jumped from burning aircraft without parachutes, their bodies freefalling towards the ocean.

Behind the initial explosions, trailing bombers pulled up,

banking left and right to avoid the carnage. I unleashed two of my external AIM-9s. The weapons curved through the air as they approached their targets and exploded, eradicating the remaining bombers.

Squadrons of Zeros changed course to assist the bombers. They approached from all directions, forming a hive around us. The Zeroes fired and bullets trailed behind our fighters.

“You’d think we just we’d kicked a beehive,” I said.

We pulled up, rising out of the soup, then Avro yelled, “Hover,” into his radio.

“Are you kidding me? No!” I yelled back.

“Trust me. I have an idea.”

I cut the throttle and pulled back on the stick, vectoring the F-35’s rear exhaust nozzle towards the hills below. Avro did the same, his gun strategically picking targets out of the sky. We had the high ground, and we were safe from diving kamikazes, but the Zeros were coming around, ready to strike.

With both F-35s in a hover, Avro kicked his rudder, rotating his F-35 180 degrees. My jet faced north, his faced south. We sat at 6000 feet, wing tip to wing tip.

Avro’s face was hidden by the mask, but he gave a sarcastic salute and said, “Clockwise rotation, decrease throttle fifty percent, on my mark.” He paused, looking for a break in the maw. “Mark!”

I stomped on the rudder and the hovering jet began to turn. Avro held formation off my port, adding forward movement to his rotation. I held in the trigger, awakening the plane’s 25mm equalizer canon. The five-barrel Gatling gun roared to life; my targeting computer allocated two 25mm guided rounds per bogey.

It was a sphere of pure chaos, as fragmented Mitsubishi Fighters and Nakajima Torpedo bombers fell from the sky. Flaming chunks of mutilated wings, bent spars, and warped steal rained down on the hills below.

I continued to hold the trigger. The firing computer chose its targets carefully, and every bullet found a home. Ordinance from the Nakajima torpedo bays rumbled like thunder as the bombs detonated in midair, or exploded in mushroom clouds as the planes careened into the hillside.

Soon, the skies over Oahu were clear. I banked away from Avro's F-35, and transitioned back into horizontal flight.

"Nice work, Johnny boy," Avro said as I came around. "Call up the Naval Air Station, frequency one hundred and thirty-two point six."

I tapped the radio controls on the touch screen, dialing into Pearl Harbor's primary military channel, and listened.

"Naval Air Station," Avro said. "This is Captain Garcia, we're coming in from Area 51, a top-secret military base, on the, ah, Big Island, inside a volcano. Permission to fly over battle ship row and clean up, over."

"This is NAS tower, who the hell are you???"

I hit the transmit button. "We're the ones who just shot down two hundred Nakajima Torpedo bombers from a Japanese strike force. Give us permission to enter Battleship row, or do you want a squadron of Zeroes flying into the *USS Arizona*? Over."

"I'll need to talk to my …"

"Thank you." Avro clicked off the mike as we switched to our private channel.

"You just love messing with Turings, don't you?"

On the deck of the *Nevada*, a marching band banged mallets on their glockenspiels; the morning's events apparently did not impact their practice schedule. Most of the sailors had no idea they were even under attack, and if they thought something was up, they probably thought it was just a drill.

I unleashed my remaining Sidewinder missiles, finishing off the remaining targets. Sailors lined the decks of the ships to watch, musicians dropping their instruments as fiery remains of the Japanese aircraft dropped into the bay. We flew the F-35s between the rows of ships including the *USS Arizona* and *USS Oklahoma*, two battleships the Japanese had sunk during the actual battle. The sailors cheered as they observed the US flags on the tails of our jets.

Between the hills, we saw the two V's of the B3 bombers, flanked by the two other F-35s piloted by Singer and Nash. The fighter battle was over so quickly that Serene and Tayler has just gotten into the air.

Serene's voice crackled over the radio. "Mighty fine show, boys, but I think it's time to let the ladies have a little fun."

"If you insist," Avro said. "Head north, and you'll find the fleet. We'll cover you."

I pulled up and away from the harbor, hitting the afterburners and letting the jet-wash create a tidal wave that washed up and over a boardwalk, soaking a platoon of camouflage clad soldiers. Avro joined me in formation over the valley, hitting the afterburners over Wheeler Air Force Base where several P-40 Warhawks rolled down the runway, ready to clean up any Zeros we may have missed.

Traveling at nearly twice the speed of sound, we met up with Tayler, Johnson, Singer and Nash five miles south of the Japanese Fleet. Amelia sat beside Commander Tayler in the cockpit of their B3 and waved as we entered formation.

"Shephard, Johnson, the battle is yours," Commander Tayler said. "Recommend you arm missiles."

The Japanese naval fleet consisted of six aircraft carriers, eleven destroyers and two battleships. Stripped naked of their defensive grid of planes, they were sitting ducks. The F-35s patrolled the surrounding area, keeping an eye out for any stray Japanese aircraft that might think to take a pot shot at our bombers.

Amelia armed the B3's extensive compliment of Harpoon anti-ship missiles. We did one pass over the Japanese carrier group, close enough to see the terrified looks on the faces of the Japanese sailors on the decks of the *Hiryu* and *Zuikaku.* Men scrambled over the deck, preparing to launch the remaining aircraft.

Tayler banked his bomber around. "Light'em up," he said.

"Pilot to bombardier," Amelia said, quoting Bugs Bunny. "Pilot to bombardier, bombs away!"

"You're not the pilot," Tayler said.

"Oh, shut up, didn't you ever watch cartoons?"

Amelia and Serene selected their targets from the displays and released the Harpoon missiles. The weapons cruised towards the enemy fleet, a ribbon of white smoke trailing behind them.

The Japanese ships were struck in sequence, destroyers and battleships ripped in half, bows and sterns shooting off, skipping across the water like cans kicked down the road. It was as if they'd been hit by a dozen speeding freight trains.

The *Hiryu* carrier's deck buckled upwards, forming a giant sphere of burning light like a mini nuclear blast. Debris fell from the sky, cannonballing into the water. With hulls breached, the ships slipped beneath the waves in a matter of seconds. By the time Serene and Tayler lined up for a final attack run, the fires were out, extinguished by the rising swells.

All was quiet on the radio until Tayler announced, "All right, kids. Let's go home."

We formed up behind Avro's F-35 like geese returning in spring. Back on the island, several forest fires burned, and firefighting soldiers rushed into the bush to put out the flames.

Smoke rose from all around Pearl Harbor, not from the American ships, but from the fractured remains of hundreds of Japanese aircraft.

We landed back at the Naval Air Station. A crowd of airmen

and mechanics met us as we hovered and landed in front of the hangars. They cheered as our canopies lifted, and kept cheering as we climbed down.

Avro walked over to me, extending a fist pump and slapping me on the back.

"Most fun I've had in years, Johnny," he said.

"I'd have to say, Kevin did well," I said.

Serene climbed down from her B3 and removed her helmet, a perplexed, yet impressed, look on her face. "I'm not sure what the commander is going to think of this little stunt, but it was unconventional. Sometimes he likes that."

The crowd parted as Tayler taxied to a halt. The whine from its engines dwindled as the aircraft shut down.

Commander Chris Tayler jogged down the steps, followed by Amelia.

"Walk with me," Tayler said, turning towards the beach. We jogged to catch up.

"Well," Avro said. "Did we pass?"

Tayler stopped, turning towards Avro. "I've been doing this simulation for years. And every one of my teams has passed," he paused, and then said, "Eventually."

"How did the others do it?" I said.

"Oh, they'd come up with some brilliant strategy, but usually not for a week or so. The exercise pushed them to their limits. It forced them to be smarter, faster. Sometimes they'd rally all the fighters into the air, hiding behind a mountain, flank the Zeroes, and hope for ten to one odds in a dogfight. Other teams won on great leadership, rallying the American troops. But hacking the VR system, programming aircraft to show up without violating the Schrodinger rule, that was truly impressive."

"So, we passed," Avro said.

"You passed, and that was some stellar flying, you two," he

said, pointing at Avro and me. "I knew you were wingmen back in Vegas, but that was something else; you have an instinct for flying that I haven't seen in a long while." Tayler paused. "That aside, we've got a long way to go before we're all acting as a cohesive unit."

"So, what's next?" I asked.

"We'll set up another simulation tomorrow. But I say we take the rest of the day off."

Tayler reset the Hawaii program, and the beach house rematerialized. Kevin showed up, back from purgatory and wearing a new T-shirt showing two F-35s, facing opposite directions; a tribute to the day's flight.

We lounged around the house drinking beer and playing cards. The beer was non- alcoholic, and the flavor-sim questionable, but psychologically, it hit the spot.

Serene Johnson studied me from across the room. She'd let her hair down, and all military stiffness and formality had departed from her posture. Avro and I had been chatting, but he politely excused himself to play a round of pool with Singer and Nash. Serene came over, leading me outside to the hammock hanging on the porch. A full moon rested near the horizon, creating flickering patterns on the water.

"You're not cocky," she said, sitting on the hammock, and signaling me to sit.

"Huh?" I replied, taking a seat. The hammock sagged under our weight, pressing our hips together in that awkward pleasurable way reminiscent of sitting beside a pretty girl on the school bus.

"Pilots are cocky," she said. "They all are, but you're not."

"I don't know what to say," I said. "You're not cocky either?"

"I'm not really a pilot, in the real world that is. I'm a solider.

But hell, I'm cocky as shit." She tilted her head as if engaged in deep inquisitive concentration, and then looked me in the eye.

"What?" I said.

"That was a joke," she said and kicked the floor, causing the hammock to swing.

"Ah," I said. "Was it?"

"Yeah, it was." Serene pivoted her torso so that she faced me and punched my arm. *Is she flirting with me?* I could never tell when I was being flirted with.

"I never liked pilots," Serene said, chin pointed down, but eyes looking up. "They never follow the rules. Never color inside the lines. That stuff gets people killed, in the air, and on the ground. You seem like a guy who colors inside the lines."

"I'm not sure if that's a compliment," I said.

"It is," she replied, twisting back into a seated position and staring out at the sea. "It's refreshing, and kind of dorky. When you're surrounded by overachieving jocks, people who flaunt everything they've got, it's nice to meet someone, well, normal."

The moon's equator bisected the horizon, and we watched it sink beneath the waves. Serene was right. I was normal. I never wanted to be, but I was. Sometimes, I felt flat, one dimensional. No matter what I did, where I went, or what I accomplished, I was still just me, an engineer who colored inside the lines. With a dead wife and son inside a broken heart.

A sensation on my leg; I glanced to see a hand on my thigh, and Serene nestled her head onto my shoulder. The sensation so real, so accurate. Serene's body was miles away, yet this metaverse connected our consciousness, like telepathy. I no longer felt one dimensional and broken; I felt alive, electric.

I thought of Marie, and immediately felt ashamed. I thought of the funeral. Would she want me to fall in love again? I concentrated on the sensations: Serene's hair tickling my neck, and her hips touching my hips. I meditated on it, clearing my thoughts of all but the moment.

I reached down, and held her hand. With the Milky Way glowing bright, we remained on the hammock, bodies pressed together, until we fell asleep.

17

Marie sat by herself in the cafeteria. It was 4 a.m. ship time and most people were still asleep in their VR suits. Her empty coffee mug hovered over the table. She reached for it, and set it down, listening to the magnetic click. This was the last meal she'd have in zero-gravity or need to fuss with aging Chinese ration packs, made doubly annoying by the fact she was eating for two. Marie looked forward to a life not confined to a VR or this cafeteria.

A countdown appeared on the holovision: "3:29 hours until deorbit." Marie took the last bite of a dry bran muffin and tossed the wrapper into the compost. She unclipped her legs from under the table, and floated back to the VR port for the very last time.

Marie walked the manicured park adjacent to her office at the Center for Genetic Diversity. The park's perfectly green grass was never more than a few inches high while flawless leaves blew in the gentle breeze. The perfection was depressing. The life she had known for the past year was only as real as this illusion

But some aspects were more real than others. Marie had purchased art for her apartment, paintings, statues, and trinkets created by real people whose names she knew. Each piece meant something; to her, they were real. She planned to print each piece once they reached Callisto.

Marie said goodbye to her office overlooking the park. She

asked James if there was an identical structure on Callisto. He said he didn't know.

A bell rang and a voice spoke over distant speakers, calling everyone to the theater. Marie took a seat near the back. Hoshi stood on the stage, her image projected onto the large screen behind her.

A growing murmur permeated the crowd until Hoshi put up a hand to silence it.

"I wanted to thank everyone here today," Hoshi said. "You've been very brave. Today we begin a new adventure, on a new world, and I couldn't think of any other people I'd rather bring along." The audience responded with a smattering of applause, before Hoshi continued, "In a moment, we're going to close the Calli simulation and prepare the ship for landing. Hold on and enjoy the ride."

The view of the theater faded.

A swift wind blew in Marie's exposed face as the giant inflatable Bigelow module began to shrink. Marie retrieved her helmet per the instruction, and placed it on her head. Her suit stiffened as the ship prepared to land. Everyone was fixed to the spacecraft by their VR ports, and with their backs to the direction of travel, as the module shrank back to its original pre-inflated size.

Thrusters fired and Marie's stomach sank as they fell toward their new home.

Marie stared out into space through the augmented window on her right. She watched as the spacecraft descended toward a barren and cratered surface.

There was a clomp, followed by a jolt, and the giant ship came to rest. Marie felt a nostalgic sensation, as if from a past life. *Gravity! Real gravity!* Not the push from a resistor suit, or the dizzying centrifugal force where the children lived, where an adult's feet spun faster than their head. Real gravity. It wasn't much, only an eighth of Earth's, but it felt significant after a year in space.

Announcements played on the ship's PA. "Please stay in

your suits. Ring airlock pressurizing," a male voice said. Eventually, the recording instructed: "You are cleared to remove your resistor suits. Please proceed to the outward hatch near the stern of the ship."

To Marie's right, several people began climbing out of their suits. A narrow stairway filled with people slowly making their way towards the exit.

Soon it was her turn. Marie hit the magnetic release on her suit, but instead of floating in place like before, the suit gave way and collapsed in a heap. Marie stood on a platform, the first time she'd actually stood in a year. Diana was on her left; she'd been quiet for most of the ride, and nodded to Marie before turning and heading to the stairs. They hadn't talked much in the past year. Most folks had exhausted their list of things to chat about months ago, and everyone was ready for a change. It was as if they'd survived a long cold winter, and this was the first day of spring.

Marie reached the stairs and gripped the railing with one hand, setting her other on her growing belly. She stepped down, testing her balance. She wasn't dizzy, thanks to the anti-nausea meds, but walking still felt strange.

The passengers moved in long single files, step pause, step pause, like getting off a passenger jet. Eventually Marie came to a hatch and entered a room resembling the inside of a grain silo. The room was grey-metallic with the word "mudroom" stenciled onto the wall. Below, smaller text read "The perfect place to shake off your space-dust." The words looked worn, probably part of an airlock used by mining colonists many years ago.

A woman in a red vest marked "VOLUNTEER" stood on a box, waving her arms and shouting, "Children over here." Marie pushed her way through the crowd and found the children lined up against the wall on the far side of the room. Mrs. Hanson and several other teachers were attempting to keep the kids separated by age.

"Branson!" Marie yelled when she spotted her son.

"Mommy!" Branson yelled back. Branson was almost four

now, and came to Marie's elbows. He had grown almost a foot since they left Earth and was also becoming quite articulate,

"What's a m-ud-room?" Branson asked pointing at the sign. Marie was often shocked at how much he'd learned on the *Mount Everest*. Their study of phonetics had several of the students reading well beyond their years.

"I don't know, honey," Marie replied. "What do you think it is?"

Branson shrugged. He wore a small canvas backpack containing one personal item: his Washington Capitals' eagle.

At the far end of the room, a circular hatch rolled open like a rock rolled from a tomb. Light streamed in and people flooded towards it.

It was a strange feeling, having arrived at their destination. There were several of those on board who didn't believe the Callisto colony existed, that they were destined to spend the rest of their lives imprisoned in the convoy, orbiting the sun, and waiting for the Earth to heal. But here they were. The Doomsdayers had been telling the truth about Callisto.

Branson tugged at Marie's arm. "Let's go, Mom, let's go!" The entrance to the Ring glowed an invigorating natural blue, the kind of light that beckoned one to go outside and experience something new.

Marie lifted her son over the threshold, into knee high grass. Birch trees swayed in a breeze. The land sloped into a valley. Rolling hills stretched for several kilometers, beyond which must be the far wall, but every horizon looked deceivingly like, well, a horizon.

They breathed the unregulated, unrecycled air. It was fresh, incredibly fresh, invigoratingly fresh. Marie took several steps into the habitat. Branson tugged on her arm, wanting to run. Was it correct to call this place a habitat? This wasn't any ordinary colony; it was a whole new world.

The new colonists spread out in all directions, some people

running, jumping, stretching their legs like they hadn't done in years. The air temperature was perfect, like a day in early fall, and Marie could smell the scent of decomposing leaves.

The trees were incredibly tall and Marie supposed the low gravity had something to do with that. Before they arrived, the architects must have flooded the area with carbon dioxide and intense artificial sunlight, accelerating plant growth. Several of the trees were ripe with fruit and in the distance a field was dotted with cattle. Apparently, the Doomsdayers had even sent an ark ahead of them.

Branson craned his head back, looking at the sky where birds circled on a breeze. In a nearby grove, bunnies scampered, looking for shelter.

"Look, Mom!" Branson said, pointing at a deer bouncing amongst the tall grass, before hopping over a natural rock wall.

They followed the crowd into the valley and Marie came to a chilling realization. Her heart skipped a beat as the implications set in. In Calli, the virtual version of Callisto, there had been houses, villages, roads, farms, and parks.

Marie looked left, and then right, down the length of the valley. Where were the villages that would house all these people? She looked back at the hatch; thousands of people from the three ships poured through, the crowd expanding into the habitat like water from a bursting dam.

Marie laid a protective hand over her belly, and gripped Branson tighter with her other hand. She realized why the villages weren't there.

They hadn't been built yet.

A young man ran out of the birch grove, yelling, “Hey! Hey! Over here, check this out! I found a sign!”

Marie and Branson headed down through the grass and over hidden rocks. Dozens of other curious people followed. A tall wooden sign towered over a natural amphitheater carved by some ancient meteor impact. Several people sat on granite rocks as they craned their necks to read the text.

“You are the first generation to arrive in this new land. The future of humanity is yours to shape. The founders, several of whom are with you this day, will be your guides, but they are not your leaders. That is for you to determine.

They ask only one thing: That this will not be a world of drones, nor a world of AI; it will be a world of natural people, and natural minds.

Don’t worry, this won’t be a society without technology, 3D printers have been located along the walls of this structure, and filament will flow like water. Utilities have been integrated but it will be up to you to build the houses. Callisto is a canvas, and you are the painters.

I suggest you begin by printing tents; it will be night soon, and there’s rain in the forecast.

God Speed.”

“What does it say, Mommy?” Branson said as they walked away, allowing others to read the sign.

“Didn’t you read it? I know you could.”

“Nah,” Branson said. “Too boring.”

“It says we need to build our own houses,” Marie said, forcing her voice to sound cheery, though she felt slightly overwhelmed by the uniqueness of this new world. Pregnant, she craved routine and stability. She also craved John. How supportive he'd been when she was expecting Branson. But now ... she felt alone with her unborn.

"Where will we build a house?" Branson said.

"Anywhere, I guess," Marie answered.

"I want to live on the top of the hill!" Branson said, pointing to the top of a large rise.

"I'll tell you what. We can sleep up there tonight. You and I can set up a tent!"

"I don't know how to make a tent," Branson said.

"C'mon then," Marie said with a smile, "I'll show you."

Marie sat on a log in front of the tent overlooking the valley as the sun set behind the not so distant horizon. Branson was inside, sound asleep on a cot. Tents littered the valley in all directions. Somewhere by the river a few hundred people were having a party. A large fire had been prepared, and a cow had been slaughtered and roasted. The smell of BBQ wafted through the air.

She was happy that there were people here who knew how to survive, how to live off the land. It was a skill no one she'd ever known possessed. On Earth, most of the meat was grown in a factory, without any animals ever being alive. This made Marie wonder, *What kind of world will my baby be born into?*

She looked up at the night sky. She knew it was a projection. Callisto was tidally locked to Jupiter, and its days were more than 400 hours long. Jupiter rose ten degrees off the western horizon, hanging in the night sky like a massive beach ball. There were several bright stars right overhead in a formation she'd never seen from Earth. But, with some simple deduction, Marie determined exactly what they were. The brightest was the real sun, the sky simulation dimming their host star to maintain a nighttime sky. There was a red planet, Mars; and three others: Venus, Mercury, and Earth.

When she awoke, Branson wandered around the tent with a stick. Marie clambered out of the tent and stretched. Their high vantage point gave them a grand view of the activities in the valley below. She watched as people gathered clothing from the printing stations on the Ring wall, bringing it back to their tents.

An engineering team wheeled several large objects from an industrial printing station. They had created an assembly area in a clearing. Large wheels were rolling out, as well as an engine block.

Marie watched as the men assembled the machine, a tractor, or more precisely, a combine. She looked into the distance, reckoning the fields from here to the water were fields of wheat. By this time tomorrow they'd have fresh bread. *A good thing to start with.*

One of the men carried a drill in his hand, its power cable snaking several hundred feet over to the nearest ring wall. With the help of another engineer, they affixed the tilling mechanism to the tractor. Marie knew this man.

"Branson, stay here," Marie said and then reached up to a grab an apple from a nearby tree. Marie gave it to Branson who took it and gave it a dirty look.

"That's breakfast," Marie said, and marched down the hill toward the engineers. "Nice tractor," she said, tapping one of the men on the shoulder. Malcom turned.

"I wanted to apologize," Marie said. "For punching you."

"Don't apologize," Malcom replied. "I betrayed you, and I feel terrible about it."

Marie cupped her growing belly and stared at him. "You prevented me from doing something stupid."

"Are you okay?" Malcom said.

"Morning sickness," Marie said, and turned around, climbing back towards Branson, who stood at the top of the ridge, still holding the apple.

By the end of the day, bulldozers were leveling ground as whining chainsaws cut through the valley, felling lumber for the first houses. City planners, trained in the Calli simulation, knew exactly what to do. Land surveys had been completed within hours and streets mapped out in yellow rope. This would become the first city on Callisto.

When the ring-shaped colony's artificial sun broke over the horizon on the third day, a row of townhomes began to take shape along the dome wall. Poly-wood frames rose from printers like treasure from Hermione's bag. The previous day's whining chainsaws were replaced by an incessant hammering that reverberated off the Ring habitat's domed roof.

Marie hadn't trained for any of the work that was going on below, and worried she would just get in the way. She kept Branson with her up on the hill, watching as he played with a few neighboring boys.

On the third evening, a man stomped his way up the hill carrying a basket of food: fresh bread, fruit, and dried meat. He was alone, but seemed happy enough, his short grey hair looking white in the evening light. This was the first time Marie had ever seen James in person, and she took note of the mild differences between the real him and his avatar. In person, his dimples were deeper as he smiled with a sincerity VR simply couldn't convey. Marie greeted him.

"So, this is Branson, eh?" he said with a smile, tipping his hat. Branson hid behind Marie.

"You should have told us there'd be nothing here," Marie said.

"Would it have made a difference?" James said. Marie didn't answer. "But, to tell you the truth, I had no idea."

"Where are the people who built this place?" Marie asked. "Will we ever meet them?"

"They went back to Earth years ago, thought they had just constructed another mining colony," James said. "I remember when my father sent up this infrastructure, the universal constructor that built this place. The Doomsdayers, like my father, didn't know when we'd actually need it."

"Ah," Marie said. "Have you talked to Hoshi?"

"I have. You're probably wondering if we should get back to work," James said. "They're going to build a Center for Genetic Diversity, but in another town."

"Another town?" Marie asked.

"Yeah, since this is going to be a major city, it will probably serve as the capital. We'll want to be far away from politics as possible."

"Was that Hoshi's decision?" Marie said.

"No, actually, that was my decision. I figured we could use some relaxation. Give civilization a chance to come together before we start meddling again."

"This civilization is going to get a whole lot bigger," Marie said, holding her belly.

"And thanks to you, we'll soon have over two thousand expecting mothers," James said. "They're already building the hospital. It will be the finest in the solar system."

"It'll be the *only* hospital in the solar system."

"Well, hang in there, and let me know if you need anything, don't be a stranger. And you need to come back to the airlock and get your watch. They've been customized for life on Callisto."

"Thanks," Marie said. "I'll do that. Then you won't have to drag your weary feet up here to tell me the news."

James laughed. "Always a pleasure." He turned with a wave, and walked back down the hill. Marie watched him go, realizing that though he was, strictly speaking, a colleague, he was also the closest person she had to a friend.

18

After Pearl Harbor, we found ourselves in the year 2043, one hundred and twenty-nine kilometers north of Tehran. A million Russian soldiers and five million citizens of the Iraqi-Iranian union were dead. Russia was losing, but for political reasons, they weren't pulling out, and so the fighting continued.

We'd been deep in simulation for almost five weeks, putting up with the extreme heat, lack of sleep, and language barriers. None of the Turings here spoke English, which had become a massive pain in the butt.

Luke and Jamaal vowed to turn our sorry asses into Marines or at least the NASA equivalent of Marines. They taught us Krav Maga, which in Hebrew literally means "contact combat", a form developed by martial artist Imi Lichtenfeld. They dragged us from our cots under cold starry skies, well before the sun rose to scold the desert with blistering heat. At mid-day our resistor suits outgassed with the smell of burnt plastic.

Locked inside our military grade VR units, there was no escape from the training; we were always "on". This Gulf War simulation wasn't like the others. When we died, there was no purgatory. Instead, we re-spawned as another solider, like agents in that ancient movie that only a sci-fi buff like me would know about: *The Matrix*. We fought alongside the local militia fighters, mainly

Shiites, who fled Tehran during the occupation. As in the prior two Gulf Wars, US special forces soldiers abandoned their traditional uniforms when imbedded with the local militias. We dressed exactly like the locals; in some cases, right down to the *keffiyeh*, or traditional head scarf. We began with four squads of twenty-four. But failed raids and snipers had decreased our number to seventy-three. I'd already been killed twice.

This night's mission would be our last in Iran. Our goal: destroy a Russian drone base. The base contained a command and control bunker which intel indicated contained several senior Russian officers. If we succeeded, resistance fighters would rise up and drive the Russians out of Iran for good.

The attack was based on a historical event. The American assault had succeeded, but no one could figure out how they did it. The base was taken, but all the soldiers were killed, and the Russians claimed it never occurred, leaving no one to tell the story. If we were successful, we'd discover a plausible explanation for what had happened there.

There was a twist. We also had to kill the Russian general leading the operation. That general was played by Commander Tayler. He'd immersed himself with the Turing computers, leading an army of soldiers against us. His mission was to take us out, making our lives hell in the process.

With Commander Tayler playing enemy, Serene was designated our new leader; she had been given the rank of captain. The six other humans on our side were lieutenants. The Turings were either NCO Marines, or local freedom fighters.

We'd arrived on location in the night, covering our Jump Jets in chameleon tarp. Stealth tents and sand colored canvas covered our equipment. From the air, we were invisible to visible light, infrared, and radar.

Captain Serene Johnson marched to the front of our canvas HQ. She wore a camouflage ball cap, with her hair bun sticking out the rear. Baggy cargo pants and a traditional wool shawl hid her figure, as well as several concealed weapons. Her face showed war paint residue from a previous operation.

"The Russian base is nearly impenetrable," Serene said, pointing a knife at a holovision display that stretched from the floor to the roof. "Its anti-shell defense system will destroy any incoming munitions. Bombardment is out of the question. Regardless, we go in tonight."

"This is cracked," Luke Singer interrupted. "Attacking a base without air superiority?"

"Lieutenant Singer, crazy is what we do," Serene replied, placing the knife in its sheath and covering it with her shawl. "I said it's 'nearly' impenetrable. We'll be going in alongside their own drones …"

"There's no way." Singer replied. "Their supply lines are guarded as well as the base itself."

Serene crossed her arms and scowled. "We're not going anywhere near their supply lines. Kevin, tell Major Singer how we're planning to get in."

Kevin was asleep; he'd been up all night doing who knows what. "Kevin!" she hissed. He stood up, stretched, yawned, and turned to the group.

"Turn off that dammed projection," he said in a groggy voice that heightened his Indian accent.

I reached over and shut off the holoscreen. With the projection off, we could see to the rear canvas wall where a strange contraption sat on a table. The device looked like a torpedo, except rendered in polygons, without a smooth surface anywhere on the design.

"We're not going to like this, are we?" I said.

Kevin smirked. "Meet Project Suckfish. I've had our printers making them all night." Kevin spread his gloved palms, and the bomb began to unfold.

"It's a wingsuit, stealth, of course. It's also a clone of the bombs the Russian drones carry." Kevin closed his palms, and the wingsuit collapsed back into its original shape. "Projection, please, John."

I was tired, and didn't immediately respond.

Kevin stared at me. I reached over and turned the projection back on.

The room shimmered and around us fish darted by. An ominous shadow approached from overhead.

"A shark," Singer said, stating the obvious.

Jamaal punched Luke in the arm. "No shit, Holmes."

"The shark is a Russian drone," Kevin said. "See that fish there, stuck to its belly? That's the remora or suckerfish. We'll clip to the drone *in flight*, just like the sucker fish."

"You've got to be kidding me. In flight?" Avro said.

Kevin nodded and used a gesture to swipe away the sea. It was replaced by an animation of a Russian drone.

"We'll land here, between the engines," Kevin explained. "You'll secure your suit to the landing boom and reel yourself into position under the aircraft."

"Seriously," Amelia said.

"The Russian drones are too stupid to detect the extra weight, and the people in the base will think they're returning with unused munitions. We should be able to get the entire squad into the hangar undetected."

"The entire squad?" Luke Singer said. "There are over seventy soldiers here!"

"Whoever makes it onto the drones makes it into the base," Kevin said.

"And those who don't?" I asked. "These things have parachutes, right?"

"No parachutes. If you miss your drone, well ..."

"Well shit," Singer said, actually sounding impressed for once.

"Are we good?" Serene asked.

"Ooh-rah," Singer replied.

Kevin took a seat and Serene stood in front of the group. "We drop at zero one hundred tonight. The Russian drones will have just finished their bombing run in Tehran."

The map reappeared behind her, and she pulled out her knife. "We rendezvous with the drones here, forty miles south of the base. Once we're in, we'll fight our way to the center of the compound. Command and Control is located underneath the palace, here." She pointed her knife at the far side of the building. "This is also where the officers, and presumably, Commander Tayler, are hiding. Once we take out the Command Center, the war is over. Everyone got it?"

"Yes, ma'am," said Nash, and Singer, accompanied by several Turing Marines.

"All right," Serene said. "Get suited up."

By zero-dark-thirty we reached 16,000 feet. The stealth Jump Jet's chameleon skin matched the blackness of the sky. The Russian drones skimmed the ground, 14,000 feet below us.

A single red bulb illuminated the cabin. As we neared the target zone, a ramp descended from the rear of the jet. We inched our way toward the edge, as a deafening wind whipped around the cabin like juice in a blender. I looked into the darkness. The other Jump Jet was just barely visible against the moonless sky.

The red light clicked off and a green one clicked on.

Jamaal Nash and Luke Singer jumped first, diving off the ramp. They extended their arms and, by extension, their wingsuits, as they slipped into the black. Serene, Keven, and I jumped next, followed by Avro and Amelia. Within seconds, all seventy-four troops dropped into the night.

Although technically a stealth airframe, the Russian drone's engines produced just enough heat to be detected by our tactical visor's IR sensors. Our suit computers amplified the signals, illuminating the drones like traffic on a nighttime freeway.

The drones flew at 270 miles per hour, a speed we could easily match. Those who jumped first aimed for the drones to the east. We worked our way west from there.

I leveled my approach as we descended below 3,000 feet. Serene flew to my left, tucking in her arms and sacrificing altitude for speed. She shot several hundred feet ahead, aiming for the lead drone. She grabbed the vertical stabilizer, swinging herself underneath the aircraft with the agility of a yoga master.

I approached my target from above, hovering for a moment and bleeding off speed to hold my altitude and then reached out to grab the drone by the tail, and missed. *Shit!*

As I tumbled end over end in the drone's jet wash, my suit tore open and the wings ripped off. I steadied myself in freefall, rolling onto my back to see Serene tucked into position on the bottom of her drone's fuselage. Her wingsuit transformed, enclosing in the chameleon shell.

I wasn't the only one who'd missed. At least twenty other bodies fell along with me.

The drones flew out of sight over a ridge. I winced, as my body exploded into the hillside. We all wore suicide vests, knowing that this was a one-way trip.

It felt like I was punched by a giant fist, the resistance suit heating to 200 degrees and attacking my skin with several hundred watts of electricity from electrodes covering every millimeter of my body. But my death lasted only moments before the program chose a new avatar.

I assumed the body of a Turing who had already activated his camo. I was now a suckerfish, stuck to the underside of a drone.

A landing gear extended inches from my face as we approached the runway. The hijacked aircraft came to a stop. I half expected the entire Russian army to be waiting for us, but the tarmac was empty.

As we passed through the hangar doors, I noticed an arm poking out of the suckerfish on the drone beside me. It was Avro's, the tattoo on his wrist a dead giveaway. He faced down, screwing a silencer on his pistol. Avro tilted his head, peering through his combat visor, taking aim from behind the forward landing gear.

Russian technicians stood watch, waiting for their drones. Silenced shots sprang from Avro's pistol and five of the Russians went down.

We dropped out from our suckers. I grabbed a magnetic-mine from my pocket, and slapped it onto the drone's fuel tank before running for cover.

Avro and I huddled behind a tool chest at the rear of the hangar. Somewhere, an alarm shrilled, and dozens of Russian troops flooded through the large hangar doors towards the drones.

Serene and a few others took cover six meters away while the enemy soldiers took up position around the drones, using them as cover. Serene raised three fingers and Avro nodded.

She folded down one finger. Then, the other. "Fire in the hole!" Serene yelled.

I hit a trigger on my wrist and the puck on my drone exploded, igniting the craft's remaining fuel. Seven drones burst apart, lobbing soldiers across the hangar and into distant walls.

Avro pointed to a door at the rear of the hangar, giving us the signal to follow him. The door led to an alley, Avro holstered his berretta and removed an MK85 rifle from its satchel. He unfolded the stock and pressed it into his shoulder, then led the way. In front of us, a solider came into view; Avro fired, and the solider exploded.

"Shit!" Avro whispered. That was one of ours. "Shit shit

shit." Our suicide vests could be detonated by a strategically placed bullet. The idea was that if we were ambushed, we'd take the enemy with us.

"Sharpen up, Avro," Serene said. "I'll lead. I'm not spending any more time in this piss-ass sim 'cause you can't hold your shit."

"GODDAMMIT," a soldier yelled from behind us. "You shot me, you son of a bitch!" It was Jamaal Nash who'd re-spawned from one of the troops with us in the hangar.

"Nash, calm the hell down, it's the fog of war, idiot," Serene whispered.

Jamaal Nash pulled his pistol from his holster and shot Avro between the eyes.

"What the hell!" I yelled.

From behind, another solider tackled Nash and grabbed him in a headlock, and started pummeling him in the gut.

It was Avro.

"Oh, for the love of God!" Serene shouted. "We need to get out of here before we all get blown to bits."

Avro picked Jamaal up by the head and threw him into the hangar wall. Three Russian soldiers turned the corner, alerted to our position by the commotion. Avro pulled out his berretta, still sporting the silencer, popping the soldiers in succession.

We jogged along a barracks, hunching down to avoid the windows as lights inside began to come on. A troop truck sped down the road, tires squealing as it rounded a corner, ready to dump its complement of soldiers into combat.

"Find cover," Serene ordered.

Across the street, a windowless concrete building bordered the main road. A lone man sprinted along its roofline. The truck stopped about ten feet from the building's northern corner.

"I think that's Kevin!" I said, pointing to the person on the roof. The running soldier wore one of our vests, but his rifle was gone, freeing his arms to punch the sky with each stride.

"Gahhhhh," yelled the man on the roof.

"That's Kevin, all right," Avro said.

Kevin jumped from the building, arcing through the air and swinging his arms in concentric circles like a six-year-old going off a diving board.

"What the ..." I said.

"*Indra Akbar*!" Kevin yelled as he fell towards the truck. His body tore through the vehicle's canvas roof; I imagine he causally introduced himself to the troops within. The truck exploded in a fiery cloud, illuminating several blocks.

"I think Kevin likes to die," Serene said.

"Yo," Kevin said as he strolled casually out of a nearby hangar, his tone hiding the immense pain of his recent death.

"Hi, Kevin," Avro and Serene mumbled in unison.

"Hey, where's Amelia?" I asked. My visor indicated the location of our troops. I saw a dozen others, their bodies outlined in green with names over their heads, but Amelia wasn't among them.

Serene put a finger to her ear. "Shepherd, do you read me?"

"You're not transmitting," I said, expecting to hear an amplified voice in my ear. "They're jamming encrypted frequencies. Switch to analog, and just don't say anything that will give away our location."

"Amelia's drone entered a different hangar. We'll find her," Avro said.

Kevin reached into a pack, retrieving an infrared pulse beacon. He reached around the building and stuck it to the wall, increasing our visor clarity.

We paused, taking note of the situation. Several of our comrades were engaged in firefights around the compound. My visor displayed fragmented data. Our troop strength indicator flashed at the bottom right. Of the fifty-one soldiers that survived our entry into the base, thirty-nine remained.

My visual overlay amplified by Kevin's beacon, I scanned the nearby buildings, and zoomed in on the palace where one of our soldiers was held hostage. *Amelia.* She winced in pain, one arm reaching across her chest to hold her shoulder. She'd been shot, and her resistance suit was probably jabbing her in the shoulder, stretching her biometrics to their limits.

Avro gave the signal to us and the troops with us. "Amelia's a hostage in the atrium," Avro said. "It looks like her suicide vest has been deactivated."

"It's a trap," I said.

"I know it's a trap," Avro said.

"I'm going to get her," Nash said, and took off running toward the palace.

"No!" Avro yelled, but as Jamaal approached the palace stairs, a sniper opened fire. Jamaal exploded in the street, his suicide vest burning the pavement.

"Well shit, that didn't work," Nash said as he respawned to become another solider behind us.

"From now on, let me do the rescuing," Avro said.

I glanced at my display: "Thirty troops remaining".

A voice with a fake Russian accent resonated through twinned bullhorns located along the palaces roofline.

"Deactivate your vests, and surrender," it said, the sound echoing off the hangers behind us. I smiled, recognizing the voice as that of our own commander, Chris Tayler.

"Kill me!" we heard Amelia yell.

"We don't have a shot!"

"You will all be killed," Tayler's voice said, tough, clearly a threat; we could detect the encouragement in it. He was probably glad to be at the end of this simulation, too.

"Why is he holding a hostage?" I said.

"He's trying to draw us out," Serene said. "We've got a lot of firepower and thirty more vests. We're about to do a lot of damage, whether we breach the compound or not."

"What now?" I asked.

"Serene, tell everyone to hold their fire," Avro said.

She did, and everyone held a defensive position. The gunfire ceased and an eerie silence settled over the base.

"It appears we have a standoff," Avro shouted across the boulevard.

"Maybe, but you're still all going to die," the voice said.

"We can still negotiate."

"No, I don't think we can," said the voice.

"Why are they holding a hostage?" I whispered to Serene. "It makes no sense."

"They're using Amelia to draw us into the open," she answered. "Wasting our resources on a rescue instead of an assault."

"Maybe we can do both."

Kevin nodded, and a serious look crossed his face. "You know what's interesting about Turing computers?" We looked at him and shrugged. "They're programmed to think they're human. The Turnings are programmed to simulate human consciousness, which itself is an illusion. Consciousness simply combines memories with feelings, giving a person the illusion of a constant stream of thought."

"Get to the point, dude," I said.

"To pass the Turing test, a computer must convince a human that it's not a computer. The way it does this is by accessing *emotions* and memories from *its* past, even if that past is artificial."

"Kevin, you're brilliant," Avro said, and got up to begin walking toward the palace.

"Avro, where are you going?" I said.

"To access the machine's emotions," he said, removing his helmet and tying back his headscarf. With his tanned skin, he looked as if he belonged in the Middle East.

Our snipers had cleared the road between us and the palace, but the Russians had formed a perimeter around us, and the palace was heavily guarded. They knew we were after Central Control and were doing everything they could to distract us.

No shots rang out as Avro stepped into the street. He reached for his shoulder, and unbuckled the suicide vest. He threw it into the street then drew his gun, and tossed it aside. He put up his hands.

"Hold your fire," said the voice. "I want to hear this."

"Before I die, I have one important question to ask her. Amelia, can you hear me?"

"I can hear you," Amelia cried in the background.

"Amelia, I've loved you ever since we met. I've loved you across the stars. In the evening, and in the morning, when we eat, and when we play."

He took two paces forward, stepping slowly towards the palace.

The snipers held their fire, listening to everything Avro had to say.

"It's too bad it ends today, but I'd trade every day from this day forward for one more chance to see you again."

"You've got ten seconds before we kill you," said the commander.

"That's all I need," Avro said. "Amelia Shepherd, will you marry me?"

"Yes, yes!" Amelia yelled, and struggled against the captor's grip.

I almost choked. The words penetrated my consciousness like a sword, tearing me from the simulation, and I forgot that we were at war. And for the briefest of moments, I felt joy.

Serene hit me on the back of the head. "Get your gun ready and point it at that door," she whispered.

"What?"

"Just do it!"

I raised my gun and Serene did the same.

"Let me see her," Avro ordered. "Let me see my fiancé, just once, then you can kill me."

Behind the wall, Amelia struggled, wiggling herself from the captor's grasp, and she stood in the palace doorway, a huge grin on her face.

"Fire," Serene ordered.

I pulled the trigger, and the bullet leapt from my gun. Time seemed to move in slow motion as the round streaked across the boulevard. The projectile penetrated through Amelia's heart and out her back, puncturing the ignition layer on her explosive vest.

The compound's lobby erupted like a volcano. Marble cinderblocks blew outward from the facade, and the entire front of the building crumbled to the ground, filling the street with dust and blocking the sniper's view.

"Move in!" Serene ordered, and we bolted toward the opening created by the blast. Bullets ripped through the air, but without clear line of sight, most of them missed.

I made it to a grand hallway, as the palace filled with smoke and dust from the explosion, and bustled with American troops. Guns blazed as we attacked the soldiers inside.

The display ticked our troop count down, "twenty-five, twenty-four, twenty-three", but it didn't matter; we were so close.

One of our soldiers leapt forward, running full speed ahead, diving over where we knew the underground bunker was. He shook off his vest, tossing it onto the floor in the middle of an atrium. A six-meter fuse connected him to the vest. He stood behind a pillar and hit the trigger. The vest detonated and the concrete below gave way, revealing a hole two meters wide in the floor.

"Grenades!" I yelled, and we all reached for our belts,

releasing the pins in fluid motion, chucking the weapons into the floor.

"Fire in the hole!" someone yelled, as a plume of dust shot from underground.

I slid to the edge, and peered over.

"Looks like a three-meter drop," I said.

Avro wrapped a rappelling line around a column, backed to the edge, and jumped. "Clear!" he yelled, as he reached the bottom.

"You four, guard the hole; everyone else, down you go," Serene said to a group of US soldiers.

I grabbed the line, and dropped in over the edge, landing in a mess of busted displays and holovisions shattered by our grenades. Dozens of bodies lay among the rubble. Along the walls sat several wounded Russian officers, their faces covered in blood. They held their hands above their heads.

The dust began to clear. Nash and Singer lit flares and tossed them toward the walls.

Nash looked pissed. "I'm fucking tired of dying," he said to a man holding his hands in front of his face. The Turing was terrified. "You know what it feels like?"

"Like this!" he said, and shot the Russian officer in the head. "Like this," he said, and shot another.

Singer walked up to him, and put his hand on this shoulder. "We have to find the commander."

"Where's Commander Tayler?" I said.

"*Gyde Ob-che-yeh*," Serene said phonetically in Russian.

"You speak Russian?" I asked. Serene just shrugged.

Our hostages remained silent, but I caught several of them looking towards an adjacent hallway. Gunfire from above us ended in an explosion; bits of ceiling rained down. Our number reduced to seven.

"This way," I said and led the way into the hall. It had only two doors. One was open. I stopped at the door and nodded to Nash. He cocked his rifle and turned the corner. Nash got off two rounds, but I heard three cracks. He took a bullet in the neck, stumbling forward and falling into the room.

Avro and Serene formed up behind me. I peeked around the corner. “Clear,” I said. Two men lay on the floor. One was dead, but the other crawled towards us. Serene turned the corner and put a bullet in the man’s head.

"Shit," she said. “We’re out of men.”

“Well it’s a good thing you’re here,” Avro said. “Try the other door.”

I nodded, fired two rounds into the handle, and kicked the door open.

We busted through to find Commander Taylor sitting behind a desk surrounded in holoscreens. He looked relaxed, probably glad the simulation was over. The commander wore the official uniform of a Russian general. He put on his cap and stood as if ready to leave. The action caught us off guard; two shots went off, and Avro and Singer dropped to the ground. We heard a revolver cocking, but Commander Tayler put up his hand. I dropped my gun.

To my left, a female officer pointed a gun at my ear. On our right, another officer held up her gun. The women were beautiful.

"So," the commander began. "You think you've made it to the end. You think you've won."

"Someone sure gets into character," Serene said. “So, you’ve just been hanging out, in this palace surrounded by broads, while we slave away in the dessert.”

“Call it a vacation.” The commander walked to me, and ripped the detonator from my vest. The fuses popped from the C4 like snaps on a child’s winter coat. He brushed some imaginary dust from my shoulder then did the same for Serene.

"I assume you've learned a few lessons along the way. You had to, to get this far."

"Sir," said one of the commander's mistresses in English. "We should execute them immediately,"

"Not yet, baby. I want to talk to them."

"What's left to learn?" I said. "We've been beaten down, built back up, learned to accept pain. We've fought, we've reconciled. What's more to learn?"

"This is war, John, a war of ideals. It's not about good and evil, it's not even about freedom. This war, and every war, is about power. It's about one man sitting at the top doing everything he can to stay there."

"And when he's dead, it's a house of cards," I said.

"Sometimes, but not always. Look around the room. All of these people, these officers, need power for their very survival. We are Russians occupying a foreign land. The people don't want us here, they hate us. Democracy will never work here because the moment we're not in power, we're dead, and we're not going to let that happen."

"You're trying to teach us one more lesson. That we're fighting a system and not just a man," I said.

"Precisely. These men and women, the officers and soldiers, were neither good nor evil. If things had been different, you could very well have been in their shoes."

"What are you going to do with us?" Serene asked.

"I would have you start over, keep going until you've learned all you can, but I think you've learned the most important lesson of all."

"And what's that?" I asked.

"Always have an ace in the hole," Tayler said and looked to his left as Kevin turned the corner and entered the room.

"Hey guys," Kevin said. "What's up?"

"Kevin, your vest,"

"What about it?

"Trigger your damned vest!"

Kevin undid a Velcro sash from his jacket, revealing the trigger for his suicide vest.

He looked at Commander Tayler and, in his most serious tone, said, "*Das vadanya,* Commander."

In a flash of heat and pain, the simulation ended.

We materialized on the front porch of the beach house in our Hawaiian T-shirts. The evening sun sat on the horizon, bisected with layers of stratocumulus silhouetted in a light pink haze. We ached with the pain of death; it would be several minutes before the pain diminished, but most of us tried to hide it; a talent we'd gotten good at after several dozen deaths. Nash collapsed and Avro helped him down into a chair.

War is hell, as General Sherman once said, but for many generations, it was a rite of passage. Our primal brains were deceived. We were a band of brothers, and sisters, albeit a very tired one.

Tayler wore a crooked smile. He was proud of us. I wanted to hit him, to punish him for the agony he'd caused us over the past weeks. "You may assign yourselves call signs now," he said.

Avro stayed with Nash, talking with him while resting a hand on his shoulder. Jamaal Nash had been willing to endure pain to save others on several occasions. Later that night, we'd give him the call sign "Gofer," for his tendency to run into danger.

Kevin stood tall, having gained respect for his technical prowess (and suicidal tendencies). We'd give him the name "Steeplechaser."

Though he was the quietest one in the group, when Luke Singer spoke, his words were profound and timely. He was given the call sign, “Sage”.

Amelia’s call sign became “Big-Guns Shepherd,” because she’d released far too much firepower on the Pacific fleet during Pearl Harbor.

Serene Johnson became “The Acrobat.”

I'm not sure what they thought of me. John Orville, a product of circumstance, not really a hero, just someone in the right place who happed not to get killed on Earth or on Mars. I entered my bedroom, looking at my reflection on a tarnished mirror, into my tired eyes. I didn’t know myself anymore. I’d changed so much that my nature no longer matched my self-image. They later gave me the “Immortal” because, as Serene said during several sims, “That bugger didn’t die?”

I splashed cold water on my face and rejoined the crew on the deck.

Avro got up from talking with Nash, and joined Amelia at the railing. They held hands and turned to face the sunset. Amelia looked back as they walked in the sand. "I'm keeping my last name," she said.

I turned to Commander Tayler. "We've got a wedding to plan," I said.

The commander looked well rested; the dessert simulation had taken a much lighter toll on him. "If you need an officiant, I'm registered."

Serene grabbed my arm, and threaded hers under it. "You'll be the best man, I assume," she said.

I looked at her and smiled. "I guess that makes you the maid of honor."

"Nah, that honor has already fallen on Kevin.”

Kevin took charge of wedding planning. We just had to put up with his constant complaining about SpaceNet. Due to our increasing distance from Earth, download speeds were slowing to mere gigabits per second. Despite the slow internet, by sunrise the next morning three custom black 1940's Chrysler Imperial limousines pulled up in front of the beach house.

The limos drove along the shore, past Pearl Harbor and into downtown Honolulu where we stopped in front of the Basilica of Our Lady of Peace. Wedding bells chimed, playing the traditional wedding song which, as Kevin pointed out, was a bit anachronistic; the song wouldn't be written until 1969.

Avro stood at the front of the sanctuary with Commander Tayler. They wore air force blues, their jackets lined with the appropriate medals. Red, white, and blue flowers arched over the altar. A congregation of hundreds of service men and women in uniform chatted amongst themselves. Kevin had set the date to December 8th, 1941; he said, "it wasn't hard to find Turings to attend a wedding for the heroes who saved Pearl Harbor."

Serene and I entered the sanctuary, arm in arm. Doves in beach-white cages cooed as we walked in through the vestibule. I took my position at Avro's side, and Serene stood across from me. Beethoven's *Ode to Joy* played softly in the background.

Kevin and Luke came in next, arm in arm with Turing supermodels in matching blue dresses. I looked at Serene who rolled her eyes.

The wedding song played from an organ. The service men and women in the audience stood, turning to face the bride.

Amelia walked down the aisle, arm in arm with Jamaal Nash. Nash exchanged a slapping man hug with Avro and kissed Amelia on the cheek before taking his place amongst the groom's men.

Commander Tayler welcomed us all together, and signaled for the congregation to take their seats.

"My superiors informed me you were a couple. I even

questioned their decision to bring you along. Relationships breed drama, and your lives have been full of it. As I recall, Amelia, you were living in a padded cell?"

Amelia blushed.

"Avro and Amelia met in the darkness of a Martian storm. They freed Amelia from immoral imprisonment, and Avro led Amelia to safety. But it wouldn't be their last adventure. They've hidden from a rogue defense force, and almost been blown up in a space ship that contained none other than a nuclear reactor. Avro and Amelia are heroes and I'm proud to join them together today by the bonds of marriage."

Tayler looked up from his notes, and said, "Amelia Stephanie Shepherd, do you take this man to be your lawful wedded husband? If so, answer 'I do.'"

"I do," Amelia repeated.

Kevin tapped me on the shoulder and whispered, "Her initials spell 'ASS' and she wants to keep her own last name?"

"Shhhh," I said, elbowing Kevin in the ribs.

The commander turned to Avro. "Avery Roberto Garcia, do you take this woman to be your lawful wedded wife? If so, answer 'I do.'"

"I do," Avro answered.

"This is the part where we would normally exchange the rings, but Kevin, I believe you have something special planned?"

Kevin nodded, reached to his wrist, tapping a button on his watch.

A flickering star danced down the aisle like Tinkerbelle chasing Peter Pan. It grew in brightness as it approached the couple, and then circled them, illuminating them in a golden haze. Avro and Amelia rose from the ground, as if gravity had vanished. They held hands, smiling, rising into a beautiful display of art and color.

Bands of twirling light circled their hands, etching colorful tattoos onto their wrists, running down their fingers like vines. The couple rotated as they rose into the air.

"I now pronounce you, husband and wife. Avro, you may kiss the bride," Tayler announced.

Avro leaned in for the kiss, bending Amelia backward. Her grown fluttered in the air and tiny stars danced around it like fairies. The congregation whooped and hollered, the deafening cheers rattling the floorboards.

As they kissed, fireworks exploded around the sanctuary, shooting from the pews.

Amelia and Avro settled back down to the ground, and turned to face the congregation.

Tayler cleared this throat. "I present to you, Mr. and Mrs. … Shepherd."

Avro turned back to look at Tayler. He just smiled and said, "Get out of here, you two."

The music returned; big band jazz appropriate for the era. Avro and Amelia ran down the aisle while the uniformed audience peppered them with rice and confetti. They shuffled down the steps of the church and hopped into a 1940 Chevrolet Special Deluxe convertible.

There is no way to consummate a marriage in VR. Despite this, we didn't see them for three whole days.

19

Clydesdale, with a population of only a few hundred people, was located twenty-four kilometers east of Newport, the capital. Natural terraces rising from rock formations ran through the town, remnants of craters the constructor had failed to whittle away. Clydesdale became known for its amalgamation of ranches, each property having stables for horses and pens for other animals. Fast growing redwoods and maples shaded the houses and barns. But nearby, a high desert-like tundra made the location ideal for grazing animals.

It was in this place that Marie now lived. The town also hosted the new Center for Genetic Diversity. As in VR, the center had a view. Its large glass windows overlooked a fifty-acre horse pasture accented with boulders like those of Stonehenge. It was far less dystopian than it was in VR.

Charles Thompson had met Diana Crane, in Cali, Marie having made the initial introduction. Marie hadn't known it at the time, but they started dating while they were still on the ship, their bodies separated by space, but their minds joined by the magic of VR. Now they lived in Clydesdale, too.

The couple had constructed their own ranch across the river. Every day, Charles would take a small boat across the channel, docking it at a wharf and walking two kilometers uphill to the CGD.

Charles had been slightly overweight when he left Earth, but was now quite fit. Whether it was Diana's insistence, or his commute, that was to blame for the transformation, Marie couldn't tell.

Marie and Branson lived in a yellow Cape Cod with a wraparound deck. Its large central chimney accented the pitched roof with gabled windows. It had been good to move into it after camping in a tent. Behind the house, James had built them a red four-stall barn that soon became home to two horses. The two empty stalls overflowed with feed and supplies. Each morning Marie would climb up a wooden ladder and toss flakes of hay down into the stalls.

It was evening on the Callisto habitat, and Marie walked her horse on a lead, while James led another. Marie's horse was a gelding named Shadow. He was mostly black with white splotches as if someone had brushed bleach over his mane. James walked a horse named Galileo, not after the Italian astronomer, but after a famous South African race horse. They were in a circular arena. Branson, who had just turned five, ran around nearby, chasing a flock of lambs.

"You know, I'm actually starting to like it here," Marie said, patting Shadow on the neck.

"It does have a kind of rustic appeal," James replied. The Ring's holographic sun penetrated the western tree line, casting pleasant shadows on the farmstead.

"The doctor is in Clydesdale full time now. A favor for me, I suppose. She said if she won't deliver my baby, no one will."

"I thought she was busy," James replied. The doctor had asked James for a horse. James informed her that no one 'owned' the animals, but that he would be happy to set up a stable for her, and let her "adopt" a horse of her own. She agreed, and the town had another happy hippophile.

Marie laughed. "Hell, yeah she is, she's been training midwives by the dozen. But remember, my baby has got a pretty good lead on Generation Hope. If he, or she, arrives on schedule that is."

"You're at thirty-nine weeks?"

Marie nodded and they walked silently for another lap of the arena until James broke the silence.

"You know, I just realized that for the first time, in a long while, I'm not afraid," he said.

"We're inches away from the vacuum of space. We're the memory of humanity. A huge weight rests on our tired shoulders. How are you not afraid?" Marie said.

"Because I've been afraid since long before Doomsday."

"What do you mean?" Marie asked.

"I always hoped my father was wrong, that Doomsday would never occur. I tried so hard not to believe him, but deep down, I knew my parents were right, and I hated them for it. At first, I distanced myself from my parents, trying to forget the inevitable. I learnt to live with my fear, and eventually, I came to peace with the idea and decided to help the Doomsdayers."

"You chose the animals that would come to Callisto?" Marie said.

"I did. I thought that would help me, I thought it would cure me of my fear, but it didn't." He paused. "Until now, here with you, I'm happy. This farm, these horses, made it all worth it, even if it cost us Earth."

"I'm glad you did what you did. Callisto wouldn't be the same without you."

The sky turned from pink to purple, and the sun disappeared. Jupiter hung high in the sky, casting eerie shadows, yet providing sufficient light to lead the horses back to the barn. They removed the horse's halters and closed the stalls before climbing the gravel path to the house.

They stopped on the deck and James rested a hand on the small of Marie's back. "Branson, go inside and get ready for bed. I'll be there soon," Marie said, opening the door for him. Branson ran inside, closing the door behind him.

The air began to cool and they stood in silence until James leaned in for a kiss. The familiar pang of grief for John shot through Marie. She put a hand on James' chest, stopping the advance.

"You know I'll always be here for you," James said.

"I know you will," Marie said with a smile. "Have a pleasant evening, James."

James took a step back, then turned, nodded, and walked home alone.

At the top of the stairs, Marie looked in at Branson through the open window. He was in the living room in his day clothes, playing with his toys.

"I told you to get ready for …" Marie clutched her stomach and sat down on the stairs.

"Branson, honey, can you get me a glass of water?" Branson did as he was told. When he left for the kitchen, Marie held her finger to her wrist. "James," she paused, letting her watch complete the connection.

"Hey," James said. "Having second— "

"James, I ... I just had a contraction."

Doctor O'Brian arrived at the house with a wheelchair, climbing the stairs and grabbing Marie under the arm. When James arrived, they led Marie down the steps and into the chair, wheeling her to the clinic between contractions. Branson trotted behind holding his Washington eagle.

"James, please wait out here with Branson," Marie said as a nurse took over. James nodded, and he sat down in the waiting room, grabbing a Dr. Seuss book from the shelf, Branson hopped into his lap. Within a few minutes of James's reading, Branson fell asleep.

Several hours later, Doctor O'Brian met them in the waiting area. "You can come in now," she said.

Branson remained asleep as James carried him into the delivery room, setting him down in a reclined rocker. James sat on a stool to the right of the bed so that he was at eye level with Marie, propped on pillows. The newborn baby blinked at its new world as it rested on Marie's chest, steel blue eyes staring into space.

Marie smiled. James had tears in his eyes. He'd been there for Marie throughout the pregnancy, and she could tell he cared deeply for her and her children.

"When Branson wakes up, tell him he has a sister," Marie said.

20

"Uncle James, Uncle James!" Branson yelled. Running out of the house, he tossed a fiberplastic glider into the air. The glider soared high in the morning breeze, twisting in the air, and catching the currents. The glider's hollow wings and fuselage gave it the properties of a very aerodynamic balloon, and it floated in the air for several minutes before landing in a rose bush several houses over.

"I made it at school. Mrs. Branch wanted us to write stories, but I wanted to print airplanes."

"Writing is just as important as building airplanes," James said.

"Yeah, but this is *way* more fun."

Lise toddled through the open door, waving an empty sippy cup at the adults. "I'll get it," James said, getting up and taking the cup from the two-year-old.

"Keep an eye on your sister, Branson," Marie said, kissing him on the head. "I love you, and I'll see you after school."

"Bye, Mommy."

Marie set her coffee on the porch railing and stretched for her morning run. She was training for a marathon in Newport that was only four weeks away. She ran down to the river and headed east. Fog hovered above still water like condensation on a mirror even as the river widened into a lake. There was a path built along the water and Marie intended to run the length of it.

After sixteen kilometers, the trail ended and Marie climbed down onto the shoreline. Jupiter's gravity induced monthly tides, and the water level rose and fell almost a meter over Callisto's twenty-eight-day orbit. After another six kilometers, Marie reached the edge of the developed area of the Callisto Ring.

She climbed up from the shoreline, stretching in a grove of tall oak trees. Something unexpected sat beyond the trees: a cabin. Near the structure, a plot of land blossomed with rows of corn. Below the corn grew what looked like carrots, cauliflower, and tobacco in rows. A robotic arm hovered over the vegetables, spraying precisely measured squirts of water on new growth. The robot arm alternated hands, like a Swiss army knife changing tools. A claw reached into the dirt, pulling out a weed.

A goat struggled against a lead fastened to a pole.

Most of Callisto's population still lived in Newport, and the others lived in towns or farms. Marie jogged up to the cabin, wondering how long it had been there. Newport's buildings had become overgrown with vines and it was beginning to look like a college campus, but this structure looked new, or at least had been cleared of vines.

Marie considered the possibility that there were others on Callisto, a remnant of the builders. It was the robot that peaked her interest; its task, weeding the garden, was one that had since reverted back to humans. She was familiar with the types of machines that could be printed here, and this robot wasn't one of them.

She walked through the grove towards the cabin, noting its proximity to the Ring wall and one of Callisto's printing stations.

"You know, you're the first person to ever find my cabin," a male voice said in an accent that sounded at first British, but then more like an actor's in an old American movie. Marie was startled at first, but the voice was calm, and she composed herself.

"Who are you?" Marie asked.

"My name is Henry Allen," the man said. "The third."

Long brown hair was pushed back behind his ears, well styled. A beard hung below his chin, as if he hadn't shaved in months.

"Marie Orville," she said. "I know of you, you were the CEO of Mars Corp. My husband had wanted to work for your company had he not gotten a job with NASA."

"Orville, you say?" he said, and his head twitched.

"Yes, that's correct," Marie said.

"I knew an Orville once. Nice man."

"A business associate?" Marie asked.

"Of sorts. Care for some coffee? I haven't spoken to anyone in months."

"Sure. Coffee would be nice."

The man stepped into his cabin. Marie followed, taking a seat in a green chair by a window. From where she sat, she could see the cabin had three rooms: a sitting area, a kitchen, and bathroom. On the far side of the sitting area, a spiral staircase led to a loft that overlooked the lower level. In the far corner, Marie noticed something she hadn't seen since before Doomsday. An Asimo robot. The drone sat in a chair, its head slouched over.

"You have a drone," Marie observed.

"Yes, I'm not as A.I. adverse as my colleagues. I used the drone to construct this cabin. He's really quite handy."

Henry Allen walked over to a kitchenette and came back moments later with a steaming mug. He placed it on a coffee table. Several newspapers lay on the table, their poly-paper sheen glistening in the light.

Marie looked at his left wrist, expecting to see one of the modified watches that all the colonists wore to communicate with each other, but he wasn't wearing one.

“You’re a Doomsdayer, aren’t you?” Marie said. It was a guess, but she felt strongly about it, and then she came to a sudden realization. “You weren’t on the spacecraft.”

“True.” H3 nodded.

“How did you get into the Ring?"

"I'll show you," H3 said. He stood, and climbed the spiral staircase to the landing above the sitting room. “C’mon.”

Marie followed him up the stairs. At the top, a steel cylinder protruded from the far wall. The wall, Marie realized, wasn't part of the cabin, but part of the Ring itself. Where the cylinder met the wall, metal twisted inward. The cylinder had punctured the Ring, like a straw through a plastic cup.

"What is it?" Marie asked, walking up to the cylinder and running a hand over the smooth surface.

"It's an airlock," H3 said. "My spaceship is on the other side."

“You cut your way in.”

“Sort of.”

“Where were you?” Marie asked. “If I had to guess, I’d say you just arrived.”

“I came from Mars.” H3 walked back down the spiral staircase.

“From Mars,” Marie said, following.

“After the Doomsday event, I attempted to save my colony.”

“From the Alliance?”

“Ah, yes, from the Alliance.” H3 settled back into a chair by a black wood stove that rested on a four by four brick pad.

“How did you escape?” Marie said, sitting in a chair across from him.

“As you may know, Marie *Orville*, I am the richest man in the solar system. I had an escape pod built into my residence. But I failed to save the colony, and a lot of people died.”

"Why are you by yourself? You have a spaceship; couldn't you have saved people?"

"I don't play games with my life, Marie. If I'd saved others, I may not have been able to save myself, then what chance would I have a rebuilding a society?" H3 said. "Sometimes the ends justify the means."

"Sometimes," Marie agreed. "But not often." She set her coffee down on a coaster. "You've given up. I don't mean to be rude, but you're an influential person, you're a leader; so what if you failed? For heaven's sake, rejoin society!"

H3 studied his guest, as if sizing her up. Then he leaned forward.

"I'm going to let you in on a little secret," H3 whispered. "But you must promise, not to tell a soul."

"I don't like secrets," Marie said.

"Oh hell, I'll tell you anyway, and then you'll understand. You'll understand why I haven't gone back to face *this* society. Mrs. Orville, the people that brought you to Callisto …" H3 let the sentence hang while he opened the door of the stove. He picked up a wooden pipe from a stand, and then began tamping fresh tobacco retrieved from a mortar resting on a knee-high mantel.

"The Doomsdayers," Marie said, to complete the phrase.

"Yes, the Doomsdayers," H3 agreed, as he lit the pipe with a match. He shook the matchstick to extinguish the flame and tossed it into the stove. "*These* Doomsdayers did something in which, I personally believe, the ends did not justify the means." H3 looked somber, as if confessing to a crime.

"What are you talking about?"

"How many ships do you think left Earth in the summer of 2071?"

"Only four that I know of," Marie answered. "The *Mount Everest*, the *Victoria*, the *Melbourne* and the *Klondike*."

“Incorrect,” H3 said, taking a puff from his pipe then leaning over to blow the smoke into the stove. A draft sucked the smoke into the chimney leaving only the pleasant scent of pipe tobacco behind.

“There were more?

“No, Marie, there were only ever three ships.”

“What are you talking about?” Marie said. Her hands began to tremble, and hot coffee splashed onto the floor. “I had friends on all four ships.” Marie thought of Lise on the *Klondike* and emotions began to flow. She felt an intense fear, and a painful cringing within, as if a doctor was about to share some terrible news.

H3 stood up, pointing his pipe at Marie. “You may not believe me when I say this, but,” he paused, “the *Klondike* never existed.”

21

Somehow Marie knew H3 was telling the truth. She realized they'd been deceived. *Hoshi, that bitch, we've all been lied to*! Marie thought. *But how far does this deception go?*

"I … I have to go," she said, getting up and stumbling toward the door. She felt drunk, wanting to heave. Marie put a hand on the doorpost, steadied herself, and then staggered outside and away from the cabin.

"Maybe you're right," H3 yelled. "Maybe I can help. You know where to find me." Marie glanced back in time to catch a cunning smile creep across the trillionaire's manipulative face.

Marie moved toward the water in a daze, as her subconscious mind processed the information, leaving her conscious mind in a funk. She trudged through the tall grass along the shore, crushing blades under her feet and leaving a trail behind her.

After several hundred paces she stopped, her subconscious having completed its silent analysis, and a conclusion reaching her conscious mind.

It all makes sense now.

Either the Doomsdayers failed to construct four ships or they knew they didn't have enough resources to support 10,000 people. In either case, they'd used her. The Doomsdayers knew the destruction of the *Klondike* would inspire a baby boom, creating a new generation when they arrived on Callisto.

Marie thought back to H3. The man seemed amiable enough, definitely well-mannered, obviously a gentleman of the old money sort. The man had ideological differences from the other Doomsdayers and his detour to Mars was understandable.

Callisto needs someone like H3, Marie reasoned.

Then she thought of James and felt sick all over again. She had to know if the man who had become her best friend was a fraud.

She jumped down onto the rocky shore, where the low tide left a gravel path, and ran until her legs ached. The further she ran, the surer she was that James knew the truth. Anger stirred her gut and she gritted her teeth, clenched her fists as she ran. She reached the place where the trail resumed and climbed onto the bank.

She ran back to her house, climbing the steps and throwing open the door.

"James!" she yelled, as the door slammed behind her, "James!"

James got up from a chair in the living room. He turned the corner, setting a book down on a shelf. Marie motioned for him follow, and she reopened the door and stepped back out onto the porch. James stepped behind her.

Marie turned, facing her foe, sweat dripping from her brow. "Lise, from the *Klondike*."

"What about her?" James said.

"You need to tell me, in one word, no explanations. One fu ... damn word. Yes or no."

"Calm down, what's the matter?"

"One word. Was Lise a real person? Yes or no."

"Marie, who did you talk to? Tell me." James took a step back, out of range of Marie's clenched fists, as if he knew…

"Get off my deck, you son of a bitch," she yelled through

tears.

"Marie!" James said, stumbling back.

"Get the hell off my deck." Marie grabbed his arm and dragged him to the stairs. She pushed him, forcing him down them.

"I can explain," James said.

"No, you can't! Good bye, James."

Marie went inside and slammed the door. She leaned against it, panting from exertion and pain. The pain went deep, a tightness in her chest and gut. She knew exactly what caused it. *I've been betrayal by my closest friend.* She straightened and tapped her watch. "Charles."

"Charles here, what's up?" said the voice from her wrist.

"We need to talk."

"James knew all along, didn't he?" Charles said, his face red with anger.

Marie nodded. They sat at Marie's kitchen table. Branson was on the floor, assembling another model airplane from the printer. Lise waddled by, holding a fiber-plastic serving spoon and banging on the cabinets as she passed.

Charles leaned forward. "Tell me about H3." The professor's anger seemed to subside, replaced by a look of scholarly curiosity.

"He said he came from Mars, having arrived right before the Alliance overran the place, just like Hoshi said."

"Okay, so now what?" Charles said.

"I think we need to go public," Marie said.

"No. For both our sakes, we can't tell anyone."

"Why?" Marie said. "People should know the truth. We can't live a world of lies, I won't do it, I just won't!"

"Marie, don't you see? They faked the *Klondike* for us! So *you and I* could do our jobs. We can't tell people about the *Klondike*; we'll incriminate the Doomsdayers, and by proxy, we'll incriminate ourselves. H3 is the only person who can take this public, and I don't think he will."

Marie looked down at the table, reached for a napkin, squeezed it into a ball, and threw it across the room. She knew he was right. They'd be putting their lives in danger, and Branson and Lise's as well.

"I want to meet him," Charles said. "I want to meet H3. Callisto needs a leader like him. If the *Klondike* was a lie, who knows what else is? Elections are in a month and I say we convince H3 to run."

"How?" Marie said. "The man is a hermit!"

"No, he's not. He's everything Hoshi's not. Hear me out. *Klondike* aside, Hoshi is too conservative. We should be sending probes or even people back to Earth. What has Hoshi done to facilitate that? Nothing. We need someone who's willing to think outside the Ring. We have more than enough resources to mount a mission back to Earth. And, now you're telling me H3 has a spaceship!"

Marie considered this for a moment. *H3 has a spaceship. This could be it, the ticket to finding out if there are still survivors on Earth. John? Don't get your hopes up,* she told herself.

"Okay," Marie said. "We'll go tomorrow."

"Perfect. Fifteen miles you say?"

Marie nodded.

"We'll take my boat."

Charles already had the boat ready to go before the sun rematerialized. When Marie appeared on the dock, he stood with one hand on a nozzle, topping off the hydrogen tanks.

"Thanks for sending Diana to babysit the children," Marie said. "They seem pretty upset that James isn't around."

"What did you tell Branson?" Charles asked.

"That sometimes adults fight, too."

Charles nodded, unscrewing the nozzle from the electrolyzer and retracting the hose into the fuel depot. An inboard engine purred to life as Charles hit the ignition. Marie settled into a white poly-plastic chair as they sped to the center of the waterway. The open-topped pleasure boat cruised comfortably at twenty-five knots as Charles adjusted the trim to drop the bow for the trip upriver.

"There it is," Marie said, pointed toward the dome's wall. Charles turned the wheel and cut the throttle. The boat's nose sank into the tide, forcing an oblong wave away from the bow. The wave traveled toward the shore unperturbed until it broke into a small white cap and splashed up onto rocks.

Charles anchored the boat fifteen feet off shore. He and Marie popped off their shoes, rolled up their pant legs, and hopped into knee deep water, sloshing their way to the shore on foot.

The hermit sat in front of his cabin smoking a stubby mahogany pipe. A subtle breeze carried the aroma down the hill, reaching Marie and Charles they trudged upward. The Asimo robot was awake now; it trimmed a waist-high hedge that was beginning to infringe on the path.

"You didn't tell me he had a drone!" Charles whispered, when they were a few dozen meters from the cabin.

"Why should that have mattered?" Marie said.

"It doesn't matter. I just never thought I'd see one again," Charles said.

H3 stood up to meet them, welcoming them with a slight

bow. He wore a plaid button-up shirt and blue jeans, as if he was the one actually working the land.

"I knew you'd be back, Mrs. Orville." H3's tone was both welcoming and mildly condescending. "I see you've brought a friend. Who might I have the pleasure of meeting?"

"Charles … Thomson," The old professor was winded from the climb, and paused to catch his breath. "Nice to meet you."

"Have a seat." There were three chairs outside. H3 gave a flick of his head to the ASIMO unit, which upon contemplating the gesture, jogged inside, returning seconds later with a tray containing two chilled glasses of lemonade.

"So, Charles, tell me about yourself," H3 said, tapping his pipe on an ashtray sitting on an upturned log. He struck a match on a nearby rock, and relit his pipe with several quick breaths.

"Well, ah, I work with Marie, at the Center for Genetic Diversity," Charles said.

"That's it?" H3 said, while taking a puff of his pipe. "You're the first people I've talked to in months." The trillionaire seemed genuinely interested, and Marie wondered if he was more extraverted than he'd led on during their first encounter, or if he was looking for information.

"I was living in Australia when they picked me up. Came here on a ship called the *Melbourne*. I have a girlfriend named Diana and I like boats, even designed my own. I like to fish, a sport banned in most places on Earth, animal cruelty they said, but you know, I think the fish enjoy a good romp on the hook sometimes."

H3 looked at Marie, while pointing his pipe at Charles. "I like him."

"How did you know I'd be back?" Marie asked.

"Because I knew you couldn't share my secret about the *Klondike*. And because you're a curious person and I have answers."

"Very astute," Marie said. "You're the one person on this

moon who didn't come on the convoy."

"How *did* you know about the *Klondike*?" Charles asked.

"Because, as Mrs. Orville here should have told you, I am, or rather was, a Doomsdayer. I helped fund their little 'mining' endeavors that made this place a reality. But, when Doomsday finally arrived, I had my own business to attend to, on Mars."

"You always were one for the grand ventures," Charles admitted. "I followed your development of the nuclear-powered ion drive. You changed the way people thought about space travel. You singlehandedly changed the way humans travel around the solar system!"

"Well, aren't you a little fan boy?" Marie said to Charles.

"It's true I changed travel for rich people anyway," H3 said. "We never could get the cost of constant acceleration drives down for everyone else; it was that blasted law that forced us to build our nuclear reactors in space."

"Politics," Charles said.

H3 winked at him as he dumped some ash in a flowerpot. Marie looked at the ashtray and then back at H3.

"It helps them grow," H3 said.

"Henry," Marie said, "what are we supposed to do? I feel like I'm living a lie."

"What she means to say," Charles clarified, "is that we want Hoshi to pay for the deception, and we need your help."

"What do you want me to do?"

"Elections are coming up," Charles began, "and we think you should run."

H3 just stared, his expression unreadable, like that of a poker player bluffing about his winning hand.

"There was a pamphlet on your table when I was last here," Marie said. "Marketing material, from one of Callisto's politicians. Which means you've been to a town and you must already be familiar with some of our issues.

"I'm very aware, actually," H3 said. "But why don't you fill me in anyway."

Charles cleared his throat. "The colony needs a highway system; it's expanding faster than our infrastructure. We've got time credit inflation, disincentivizing people from saving money. There are animals everywhere, and there's a vocal minority lobbying for a zoo."

"Those are trivialities, Charles, tell me about a real issue. A juicy one, one that we can win on."

"A reconnaissance mission to Earth," Marie said. "I want to know everything about the Doomsday mission, no more secrets."

"Defense of the colony," Charles said. "It's worked for a generation of politicians. That and lower taxes."

"Now that's more like it," H3 said, setting his pipe down on the stump. "I think you've proposed a viable platform: economic progress, and defense. I'll tell you what. I'll run on one condition."

"What's that?" Marie asked.

"That you help me, Marie, help me every step of the way," H3 said. "I'd like you to be my personal assistant. Charles, any help you can provide would be greatly appreciated."

Marie glanced at Charles. He looked almost giddy to be asked to help one of his heroes.

The temperature outside dropped and it began to rain.

"Come inside, please; I'll make some tea," H3 said. They went inside and took a seat. Charles reached over and turned up a heating element.

"Does Hoshi know you're here?" Marie asked.

"Oh, I don't think so," H3 said, coming back from the kitchen. "My ship wouldn't have shown up on radar … or LIDAR for that matter, and besides, this outpost isn't very technical."

"What about James?" Marie asked. "He probably suspected

something when I told him I knew about the *Klondike*."

"Ah yes, Harrison's boy. I know his father."

"You mean you *knew* his father. He didn't come with the convoy," Charles said.

"Oh, yes of course," H3 said, without a hint of grief. "It doesn't bother me if Hoshi knows I'm here. She's going to find out eventually. In the meantime, tell me how things work around here."

"Callisto is ruled by a represented government," Marie began. "There are twenty-five electorates, basically one for each district or town. Each district elects one representative. Newport, the capital, being the largest town, elects five."

H3 nodded. "Yes, yes a very good choice in governmental structure."

"Every two years, citizens vote for their local representatives. A week later, the representatives elect an executive committee, which includes the election of a consul."

"Apparently, the citizens found the word 'president' too dictatorial," Charles added.

"Perfect, perfect. Help me get elected as a representative, and I'll handle the rest," H3 said. "Which district am I a part of?"

"You're in ours. Clydesdale is the closest town," Marie said.

"A man named James Bekker is our current representative," Charles said. "So far, no one is planning to run against him. The process is simple; there's a town hall followed immediately after by the elections. I say you show up, and steal the show."

The rain slowed to a trickle and the holographic sun began to peek through the clouds.

H3 nodded towards the door. "I'll meet you in Clydesdale, Marie."

"Do you need us to pick you up?" she asked.

"No that's alright, I'll meet you there," H3 replied, and pointed to the robot charging in the corner.

"Charles, why don't you come back this afternoon? You and Asimo can help me build a boat."

22

Clydesdale Elementary School's multipurpose room had never been so packed. The chamber, where the town hall was held, had been wreathed in carvings of historical relics from Earth. An Eiffel Tower climbed one wall while a Japanese pyramid-city lined the other. The two structures met overhead like swords locked in battle.

Multicolored chairs were arranged in two sections, as in a wedding chapel, each section eight rows across. Marie, Charles, and H3 sat at the back. H3 wore a baseball cap and a flannel shirt. He'd shaved his beard, but left a few days' scruff, and looked more like a farmer than a trillionaire. He reached into his pocket and pulled out his glasses, a concession to style he wouldn't part with, even while trying to blend in.

James stepped up to a wooden podium, and spoke into a small microphone rising from its center. "Welcome everyone, welcome," he said. The murmuring room faded into silence. One person coughed, a sound which echoed off of the high ceilings.

"It's been an honor serving as your representative these last two years. Citizens of Clydesdale, you have created a town we can be truly proud of, a town where we've found joy and tranquility, despite our turbulent past. Please, give yourselves a round of applause." The residents loved their town, and after the applause subsided, he continued, "I know you'd like to skip right to the election, but per our new constitution, I'd like to open the floor to final nominations."

People looked around, waiting for someone to raise their hand. James had been a respectable representative. Clydesdale was well designed and well managed. They'd taken care to design each structure tastefully and people would often visit from Newport to taste Clydesdale's apple ciders, and ride a horse and buggy around the trails through the woods.

James leaned toward the microphone. "Alright folks, if no one—" he was cut off.

"I nominate Henry Allen!" Charles yelled from the back of the room.

At first, the people thought the name was a coincidence and not everyone turned to look. Henry Allen was well known on Earth, and was often likened to Howard Hughes, or Elon Musk. There was no way, Henry Allen *the Third* lived in Clydesdale.

"Seconded!" Marie yelled.

"Well," James said, taking a step back, but leaning forward to speak one last sentence into the microphone. "Would Henry Allen like to join me at the podium?"

H3 stood, pacing confidently to the front of the room, and then jogged up the two steps and joined James on the platform. The two men shook hands, and H3 beamed as if there was no place in the solar system he'd rather be. H3 took off his ball cap, and ran a hand through his brown hair that retained a meticulous style despite its recent imprisonment under the hat. James recognized him, and frowned.

"I'd like to say a few words to the people of our fine town," H3 said, placing a hand on James's shoulder, his words half picked up by the microphone, half resonating off the walls.

"Of course," James said, and stood back.

H3 centered himself on the podium. "My name is Henry Allen," he paused to place the ball cap on a shelf within the pulpit then said, “the Third.”

There was a commotion as people whispered to each other.

H3 looked around the room, took a deep breath, and continued, “As you can tell, I’m new here. And not just to Clydesdale, but to Callisto. My spacecraft is docked right outside your fine town.”

More commotion. H3 smiled and put up his hand before continuing. “You probably want to know my story.” He rested a hand on each side of the podium and leaned forward, eyes scanning the now silent crowd. “After the impact, I went to Mars to save my colony. I wanted to make peace, but I failed and barely escaped with my life.”

He paused, letting people’s imaginations fill in any gaps. “I’m familiar with your local issues, the stray animals, the zoning rules, time-credit inflation and all the gobble-dy-gook, but that’s not why Dr. Thomson and Dr. Orville nominated me. They nominated me not just to represent Clydesdale, but for the betterment of the entire colony. We’ve got bigger issues than just this town. We live on a world where only four inches of flex-glass separate us from the vacuum of space, and Jupiter’s radiation. It’s a dangerous universe out there. We need leadership willing to face those challenges head on.”

“Are you a Doomsdayer?” someone yelled.

“Yes, I was,” H3 said. “The wealthiest of them all actually, which is why I went to Mars, to be with my people. The residents of the Harmony Colony were my employees. I was responsible for them. It was only right for me to join them.” He paused, as if giving his former employees a moment of silence. “But as I said, I failed. The Alliance ships entered Mars’ orbit with their detachment of ‘survivors’.” H3 paused to make air quotes. “These weren’t common people like yourselves, but the CA’s entire elite class, and their guards! They sent a scouting force to the surface and blew up one of our domes, killing thousands.”

H3 leaned forward, his lips millimeters from the microphone. "Here's the deal, citizens of Clydesdale. The Communist Alliance won't be content to stay on Mars. Mars has no universal constructor. It has overcrowded domes. They're going to want to come here, and we'd better be ready for them."

Hoshi had told them to fear the AI, taught them to fear space. Now, H3 was giving them a new thing to fear, and he would use that to his advantage. He let the commotion grow then silenced it with few taps on the microphone.

"When you vote, don't just think of Clydesdale, think of who will keep you safe. We're in space, people, and there's no place to hide."

Dozens of people spoke at once, raising their hands and shouting. James stepped forward, and yelled, "One question at a time, please." He held up his arm, until the people in the room who wanted to speak did the same.

"I want specifics," said a woman.

We'll keep tabs on the Alliance," H3 said. "We'll send reconnaissance ships to spy on Mars. And if they try to take our colony, by God we'll defend it! A vote for me is a vote for your very survival!"

"This colony is huge!" a man shouted. "We should welcome all humans here with open arms!"

"Why yes of course we *could*, and in any other situation, of course we *should*," H3 said, "but you need to understand the reality of the current situation. The Communists will be happy to share this fine colony with you, but on their terms. There'll be no more democracy, and no more freedom. You and your children will serve their flag. Your values will be their values, or you will die. Don't you remember what happened on Earth, when the Alliance republics cut themselves off like North Korea before the Great Reunion? Defectors watched their family's deaths broadcast over the internet. Towns were burned to the ground! You don't need to be reminded of what the Alliance is capable of."

H3 breathed heavily before slouching back, his rant exhausting at least some of his energy. He stepped off the podium, walking down the aisle with the confidence of a CEO. A murmuring returned as the people of Clydesdale contemplated their fate under a cruel dictatorship.

James returned to the podium. "I don't quite know how to follow that," he said. "Are there any other orders of business?"

"Open the polls," someone yelled.

James glanced at a clock on the wall, looking for some reason to delay, but there was none. "Ah, yes, please, ah, go ahead." He nodded to a clerk in charge of overseeing the election.

Around the room watches buzzed, and citizens stared at their screens and scrolled through the ballet. Marie did the same. H3's name was listed at the bottom, having been added at the last moment. She took a breath, still stewing in anger over the truths she knew James held secret. She tapped H3's name and confirmed her vote.

H3 himself pulled out a gold pocket watch, on a six-inch silver chain, and watched something on its screen for the next five minutes as votes poured in. It was as if he was so sure of winning, he didn't even bother to watch the tally of votes appearing on the projection behind the podium.

As the last votes trickled in, the result was uncontestable: Henry Allen the Third would represent Clydesdale.

H3 stood at the helm on the top-deck of the pleasure craft. Marie stood at his side, the wind blowing the curls from her face. The yacht was ornately decorated with fiber-wood paneling, while tables and chairs rose from a bamboo floor. Aerodynamic awnings covered the cockpit, made from interlocking sails. This gave the boat the look of a sailing ship whose mast had folded in on itself, as if designed for supersonic flight.

They berthed at Newport marina and walked toward town on cobblestone streets. Several hundred citizens whizzed by on hydrogen Vespas while others peddled fixed-gear bicycles.

Marie and H3 strode to Government House, subjectively the nicest building in Newport. The building was made of stone, taking two years to build, versus the days it took to print a typical structure. They stepped into an elaborate lobby with a red carpet running up parallel staircases, and climbed to the session chamber located on the second floor.

Double doors opened to an auditorium, its walls decorated with portraits of old world leaders. A domed ceiling of Renaissance art and wood carvings rose fifteen meters above the floor. Rows of desks on arching levels surrounded an oval platform. At the front of the room, three executive desks rested on elevated pedestals. A holovision behind the central desk, Hoshi's, displayed the agenda for the day's council meeting.

Hoshi sat alone as Marie and H3 entered. The reserved woman leaned over a leather notebook, making notes on her tablet.

"You're early," Hoshi said, without looking up. "Session does not begin until nine."

"I'm always early," H3 said. "It helps me finish before others have begun."

"You two clearly know each other," Marie stated.

Hoshi stopped writing, but did not look up. "Your theory about an Alliance invasion precedes you." As if at great expense, Hoshi lifted her gaze.

"I learned from the master," H3 said.

The other representatives filed into the room, locking eyes with H3 as they took their seats. A camera broadcasted the meeting to any interested colonists, while a holovision off to the side, projected a mirror image of the chamber. Marie looked at HV, and was thankful she appeared as one among many.

At 9 a.m., Hoshi banged a gavel, silencing the room with one loud crack. "The meeting will come to order." She scanned the room, and then said, "Are there any changes to the minutes from our last meeting?"

Marie looked over at H3, who was rolling his eyes.

"If not, I'd like to welcome, and congratulate, all the new members to the council; please give them a round of applause." Hoshi let the applause ring for no longer than five seconds, before hitting the gavel again. Marie wondered how this woman maintained power. Apparently, the citizens of her home district, Newport, respected her efficiency.

"First order of business, establishment of zoning rules for Newport's waterfront—"

"First order of business," H3 interrupted. "I'd like to put forth a motion." His voice echoed across the walls, followed by a silence that made Marie hold her breath. Hoshi's eyes burned with anger; she reached for her gavel, then retracted her hand for there was no need of a gavel in a silent room.

There was a reason no one broke the silence. They were expecting a speech.

H3 scanned the room. "I've read this colony's laws, laws that were written in this very room, laws that were *supposed to* ensure the *good* governance of this civilization. However, these *laws* are insufficient."

"This can wait, Henry," Hoshi said.

"No, the zoning can wait," H3 said. "I'd like to work as soon as possible to provide our citizens with peace, security, and prosperity. I was elected to insure the good government of this civilization. And that includes its safety."

"You haven't even been sworn in," Hoshi said.

“A technicality,” H3 said. “In the interest of expediting matters, I recommend we suspend these archaic formalities, indefinitely.”

"Agreed," said a man to Marie’s left. “Motion to skip the ceremonies. We’ve got more important things to discuss.”

The name plate on his desk read “Sherman T. Matthews, Lakeview District.” He continued, “I heard what he said in Clydesdale; if he’s right, we’ve got no time to waste.”

“Seconded,” said a woman whose desk plate read “Sarah Collins, Florence District.”

“Well congratulations, Henry,” Hoshi said. “You’ve overthrown one of the only traditions we have, not to mention Robert’s rules of order. Are you going to keep interrupting, or can we get on with new business?”

“New business,” H3 said. “Motion to let me have the floor.”

“Approved!” Sherman yelled.

“Seconded!” said Sarah Collins from Florence District.

There was a commotion, and Hoshi banged her gavel three times. “That’s not how this works!” she said.

“It is now,” H3 said, letting his gaze wander, making eye contact with the other representatives. “I’d like to direct your attention to the display for a moment. I want to show you something.”

H3 retrieved something from his pocket. He held the object up for the room to see. It was a gold pocket watch, connected by a six-inch silver chain that formed a loop around his thumb. He placed the watch on his desk, which lit up as it received data from the device. With a sweeping gesture, H3 transferred a video to the holovision located above Hoshi’s head. Hoshi directed her eyes to her desk which mirrored the presentation.

“This is video from a security camera on Mars,” H3 said.

The projection showed a wide-angle view of a pavilion.

Debris from some unknown source exploded onto the floor, covering mosaic tiles with a crest that read "Alamo". Two crushed police cars sat with punctured tires and crushed roofs. Suddenly, bullets ripped into the vehicles, and one of the doors fell off. The scene was so vivid, it took Marie a moment to realize there was no sound.

Police officers took cover behind their cars, while civilians ran diagonally to the line of fire, several of them taking hits. They ran toward a set of doors held open by a rectangular wedge. The words "Tram Station" were embedded in marble above the door.

A police officer went down, his weapon sliding several feet along the ground. Even with the officer clearly incapacitated, bullets continued to score the surrounding floor and walls, as if the attackers acted from vengeance.

Marie checked the community feed. The view count had spiked north of 4,000. Over half the colony was watching.

"Look here, and here," H3 said, tapping his desk, highlighting one quadrant and zooming in. Soldiers in military spacesuits took up position behind pillars, firing shots at the police and civilians running through the scene.

"Those officers are using cross bows! The civilians have spears and bats," Sherman said, pointing at the display. "This is unreal!"

"There were no guns on Mars," H3 said. "Not until the Alliance arrived. We printed these defensive weapons from primitive designs before they shut off the power."

"This has to be fake!" someone yelled.

"Run a VR detector. I assure you, this video has not been tampered with."

"It checks out," said a young man seated next to the Lake View Representative.

On the screen, a barrier rose, sealing off the pavilion from the rest of the colony. A solider in a spacesuit tore away the panel that had propped the Tram Station door. Seconds later, the remaining civilians held their necks, and fell to the ground.

"The Alliance suffocated them," H3 said. "All of them. The attacking force wore spacesuits. When they were done, all they had to do was clean up the mess. The colony was theirs. As long as you didn't resist, you were allowed to live. I stayed on Mars for a while, trying to negotiate from my safe room. It was useless. I used my escape pod to rendezvous with my spaceship, hidden on Phobos."

Marie glanced once more at the feed. 6,000 people had tuned in.

H3 removed his watch from the desk and put it back in his pocket. "Now, as I'm sure you realize, one thing is clear ..." he paused, half expecting someone to complete his sentence, but the room maintained its awestruck silence.

"And what's that?" Hoshi said, feigning boredom, the only person in the room seemingly unmoved by H3's presentation.

H3 stood up, and looked around. "Callisto needs an army."

"Absolutely not!" Hoshi shot back.

The room erupted in shouts. "Why not?" several representatives shouted at once.

"*Because the Doomsdayers were pacifists*," someone taunted.

"They're not *all* Doomsdayers!" another representative shot back.

Hoshi banged her gavel until the room quieted down, then spoke, "The founders of this colony would never have agreed to this."

"And by founders, you mean yourself!" said the representative from Lakeview, before continuing, "There are a thousand reasons we need an army, and your only objection is, 'the founders would never agree'?" he paused. "Motion for a referendum on the creation of an army."

"Seconded," someone shouted.

The room went silent.

"How are we supposed to pay for this army?" said a small old man from the back of the room. "Last I checked, we can't print time credits."

H3 turned to face the small man, and looked at him over his glasses. "You collect taxes, don't you?" he said. "Treasury Secretary, how much are our taxes?"

"Flat tax, ten percent time credit," said a middle-aged lady dressed like an accountant. "One hour per day for each working citizen."

"Make it fifteen percent," H3 said, doing some mental math, a skill refined by all the best CEOs. "We have five thousand working people here. The extra half hour will more than fund our little army."

"You make it sound simple," Hoshi said.

"This is war. It is simple," H3 said, calling up a spreadsheet application on the primary display. "We'll pay our *volunteer* soldiers two credit-hours per day." He paused to add in a few figures, with a steady hand. His fingers twitched like typewriter arms smacking inked letters onto a fresh manuscript. "Budget another credit hour for miscellaneous expenses ... A standing army of four hundred soldiers should cost us less than," he paused to finish his calculations. The fully itemized entire budget appeared on the screen. "... twelve hundred credit hours per day."

"Your economic analysis is naive," Sherman said. "War has never been this—cheap!"

"I have simply modified the economics to fit our unique situation," H3 said, "We're the underdogs here, like rebels fighting the empire." H3 sat down, having given the council members much to contemplate. The room erupted in murmurs, and once more Hoshi was forced to use her gavel.

"We'll take a three-hour recess while you prepare the statutes," Hoshi said.

The council members stood and began moving toward the

exits.

"An army requires soldiers," Marie said. "You're going to have to recruit."

"Nothing unifies a people like the prospect of war," H3 said with a smile.

"I expect our enlistment numbers will be overwhelming."

⚡

The council reconvened after the recess. "Are the statues complete?" Hoshi asked the room.

"Statues are complete, and the publication is ready for distribution," said the secretary, a young man seated in the corner of the room.

"Let them vote," said Hoshi.

"Opening the polls," said the secretary, glancing up in anticipation of a response from the room. A moment later, everyone's watches chimed in unison. "Poll's open," he confirmed.

Around Callisto, the population stopped what it was doing, with watches overriding any "Do not disturb" modes.

Democracy would not be interrupted.

Marie extracted the screen from her wristwatch and scrolled through the proposition labeled:

"Referendum on the establishment of the Callisto Defense Force."

At the end of the document, which was perhaps a dozen pages long, there were three buttons: green, which was marked "yea", a red box marked "nay", and grey box marked "abstain". Marie looked around the room and saw the others casting their votes.

Did she fully understand what they were about to do? Did

anyone? Marie thought back over past wars. She was at George Washington University when Gulf War Three raged in the Middle East. The school had taken in several hundred refugee students, many of whom were the only ones left of their families.

She thought of pacifist societies, and what happened to them. The Acadians in Nova Scotia, during the war between the England and France, were French-speaking farmers who refused to swear loyalty to either side. The Acadians were rounded up like cattle, imprisoned in ships for months, and then dumped as far south as Louisiana.

Marie scrolled through the fine print, words that H3 had dictated only minutes earlier. H3 had been meticulous in his appeal, itemizing command structures, logistics, and hardware in great detail; the entire proposition exceeded 5,000 words. As the votes rolled in, she concluded that few citizens had read the entire document.

Several pages in, hidden in the middle of a sub-paragraph, there was a note labeled "Security of the government". H3 had requested personal security guards for select members of the council. *What is he afraid of?*

The screen behind Hoshi displayed three columns: yea, nay, abstain. The "yea" column climbed almost immediately to eighty percent.

Marie slid the screen back into her watch; her vote wouldn't make a difference. The people would have their army. By the end of the day, 400 people had signed up for the CDF.

23

"Hey, Marie!" cried a male voice.

Marie turned, having just stepped out of a coffee shop in Newport, and saw Malcom jogging toward her. "You're working for H3, right?"

"Yeah, why?" Marie said, still skeptical of the man who'd deceived her back on Earth. She took a small sip of coffee, testing its temperature.

"We've made some progress. I mean, we're going to contact Earth, well sort of." He took a break. "The moon, we're going to call the moon, Earth's moon."

"Slow down," Marie said. "What are you talking about? I thought all the radio frequencies were jammed." She turned to face him.

"The frequencies are only jammed around Earth. But that's not the reason we haven't sent out signal. If we send out a radio signal, any radio signal, we'll be broadcasting to the entire solar system."

"And that's bad, why?" Marie said.

"We don't know what the Alliance, or the AI, knows about us, and we want to keep *new* information to a minimum."

"So, you're looking for permission to send a signal? Good

luck with that."

"No, no that's not it at all," Malcom said. "According to Hoshi, the Doomsdayers left a small group of people on the moon. They're the ones in charge of quarantining Earth. But we've never been in contact with them, as there's never been a need."

"Is there a need now?" Marie asked.

"Yes, for those of us who want an update on Earth, they're the best people to ask."

"Okay, but you just said you can't send them a message"

"No, I said the radio frequencies are jammed. We can still send a message optically."

"Optically?"

"Using lasers," Malcom replied. "Usually, you send a regular radio pulse first, to let the receiver know a laser transmission is pending. Once they receive that signal, they align a detector in the direction of the pulse."

"So, they have to be listening," Marie said. "Are they listening?"

"We don't know. Hoshi *says* they're not, but what if they are? Huey and I are planning to give it a try. We'll send the message out blind, in hopes someone is listening."

"Have you sent any messages yet?"

"We're still refining the coordinates."

"How long?"

"An hour, at most," Malcom said. "I was hoping H3 would be supportive."

"You're worried that Hoshi will try to stop you," Marie said.

"Exactly. That's why we've waited until the last minute to tell anyone."

"This is the most progress anyone's made toward contacting Earth," Marie said. A tiny flutter of hope brushed the inside of her ribs. *Earth. John.*

“This is the *only* progress,” Malcom said. “We’re set up in the mudroom, just inside the airlock. Come as soon as you can.”

Malcom turned, and walked toward the habitat’s northern wall.

Marie turned, crossed the street, and climbed the stairs of the capitol building. She walked to H3’s office. His new security detail greeted her at the door. They wore tan uniforms and side arms. H3 had chosen the toughest and biggest recruits for his private security, and it made Marie uncomfortable.

H3 was in his office. He looked up from his desk as she walked in. They were getting used to working together, and he no longer smiled. “Yes?” he said.

“An acquaintance of mine is going to send a message to the lunar colony. He thought you’d like to know.”

“Really?” H3 said. “Where is this acquaintance?”

“They’ve set up a communication station at the airlock. They call it ‘the mudroom’”

H3 stood up, and buttoned the top button on his blazer. He marched to the door, motioning to Marie to follow him.

“Gentlemen,” H3 said to his security detail. “Could one of you please get me a coffee? *Venti*, please.”

Malcom had set up a receiving station in a loft near the ceiling. Huey was there as well, up on a ladder feeding some cable into the panel in the roof.

“There’s a transmitter outside, right above us, and we can control it from here,” Malcom said. “Most of the comm unit was gutted when we arrived; whoever was here before us took it with them when they left.”

Huey whistled, getting everyone’s attention, and then gave a thumbs-up to Malcom.

"Don't expect a quick reply."

"We're aware of the communications delay," H3 said, taking a sip from the large coffee mug he was holding. "Carry on."

H3's guards paced around the room, inspecting the equipment, but otherwise staying out of the way.

Malcom picked up a microphone from the table, and pressed the transmit button. "Lunar station, lunar station, this is Callisto calling. We are survivors from Earth, if anyone can hear this, please reply. Direct an optical transmission to Callisto, I repeat, the Jovian moon, Callisto." He released the transmit button and took a breath.

"Turn it off," H3 said. Malcom looked at him, confused.

"What?" Marie said.

"I've decided not to let anyone continue with the communication project," H3 said.

"Why?" Marie asked.

"It's too risky," H3 replied.

"But what about contacting Earth?"

"We'll make contact with Earth via a reconnaissance mission when we're ready. I'll not risk the lives of those here in the colony."

"Screw that," Malcom said, then pressed the transmitter.

"This is Callisto station, I may not have a lot of time, a man named H—"

H3 reached over and switched off the transmitter.

"Three," Malcom finished. "What the hell!" He stood and faced H3 as if ready to throw a punch.

"You can't be serious?" Marie said.

"I'm dead serious," H3 said. "Matthews?" He gestured to one of his guards.

"Yes, sir?" the man responded.

"Take this equipment and dump it in the nearest recycolizer."

"Now listen here," Malcom said. "We've worked hard on this, and I won't let you destroy everything we've worked for."

The guards took a step closer to Malcom.

Malcom continued, "You don't have the authority to stop us. If need be, give the colonists a chance to vote. It should be up to them as to whether this risk is justified."

"I don't think you understand," H3 said with cold authority. "Power lies with those who take action." H3 took his cup, pulled off the lid, and poured the entire contents all over the equipment.

Popping sounds emanated from the motherboard as connections short circuited. The screen on the computer display flickered, and went blank.

"You son of a bitch," Malcom yelled, and took a swing at H3. He stepped back in one easy motion, face inscrutable, eyes narrowed. Matthews reached out, and grabbed Malcom's fist, twisting his arm behind his back. Huey, who was now down off the ladder, ran over to help his friend, but H3's other guard grabbed him by the collar and forced him over the table.

"Henry!" Marie cried, grabbing H3 by the shoulder. It felt hard as stone, and did not bend at her touch.

H3 growled, "I won't stand for unnecessary violence in my colony."

"What are you doing?" Marie shrieked.

H3 breathed heavily through his nose, and stared at the floor for a moment, as if trying to rein himself in. When he straightened, the mask of his face was smooth and urbane once more. He nodded at Marie and motioned to his guards to release Malcom and Huey.

"I apologize," H3 said in a tone now lacking any hint of menace. "I overreacted. I don't respond well to threats of violence."

Marie stared at her boss uncertainly, a chill uncoiling in her stomach. *Who is this guy, really?*

24

A half billion miles from the sun, Jupiter's gravity began reeling us in like a fish on a line. Our VASRM engines fought the pull with steady persistence as we arced into a trans-Jovian-injection orbit.

All of this happened without human intervention. The ships followed a trajectory planned weeks ago. We were caught off guard when Taylor recommended we switch to augmented reality. The thought of leaving VR was strange after so much time inside. I now understood why people got addicted. Virtual Reality was utopia. The human mind handled this like a drug, wanting more and more immersion. It had been our home for the past several weeks, and I knew we'd miss it once it was gone.

I brought up the spacecraft's main menu, and made the switch to AR. Activating the JJ's external cameras granted a 360-degree view of space.

Jupiter hung to my right like a giant marble. At over sixteen million kilometers distant, I could practically reach out and touch it. The planet fit between my thumb and forefinger; its four largest moons orbited nearby like pearls on an invisible necklace.

Io orbited nearest to the giant gas planet. Volcanoes on its surface accented a sea of swirling pigment, where oranges, yellows, and pinks mixed together like paint on a pallet.

Europa floated further out; hexagonal ice plates covered its surface in a mosaic. Below, an underground sea swirled with twice the water of Earth's oceans. Twenty years ago, NASA planned to send a submersible to Europa's watery depths, but the mission was never funded. For the past two decades, only mining corporations had ventured this far into space.

The giant grey moon of Ganymede was next, the largest of Jupiter's moons. It looked quite a lot like Earth's moon, except for the Texas-sized ice cap accenting its southern hemisphere like a tattoo.

I had to crank my neck far to the right to see Callisto, the smallest of the large moons. At this distance, it looked like a bright star moving against a fixed background. We were closing in.

"Orbital insertion in progress," Tayler radioed. My spacecraft yawed, directing its thrusters for the maneuver.

Callisto grew as we curved in to meet it. Indicator lights flashed on my AR console, but I ignored them. The insertion had been preprogramed before we'd even left the moon.

My little ship rumbled as chemical thrusters came online, augmenting the ion engines that burned at high specific-impulse for the duration of our journey. The chemical thrusters would consume more hydrazine in the next ten minutes than had the nuclear-powered xenon ion engines during the entire trip to Jupiter.

Callisto grew larger until we could see every crater. Its surface was a dirty brown from pure iron oxide, the element that made this moon such a valuable target. The craters were white, proof that underneath the surface, much of the world was composed primarily of pure water ice, the lifeblood of any manned colony.

Thrusters powered down as the Jupiter Jumpers entered orbit sixty-four kilometers above the surface. We circumnavigated the equator, passing into Callisto's shadow. As the ships wrapped around the small world, we witnessed our first Callisto sunrise, crossing the terminator into day.

"There it is!" I heard Amelia shout.

"Where?" I asked.

"See that row of white craters? You can make out the Ring along the equator."

I put my hands up as if framing a photograph, then pulled them apart. The computer recognized the gesture and zoomed in where Amelia had mentioned.

Even with magnification, the Callisto Ring was barely visible, but two sets of constructor tracks gave it away. Spaced three kilometers apart, the rails stretched out to the horizon. The Ring colony lay in between. Its roof had been colored to blend with the surroundings, right down to the white craters.

We'd continued for 1,609 kilometers over the Ring when the tracks stopped. Then I saw it, the device that made it all possible: The Universal Constructor.

"It looks abandoned," I said.

"It's cool," Kevin said. "I mean literally, it's cold. Aside from the decaying uranium in the reactors, thermal scans come up negative. If there were people inside the machine, we'd know."

"If it's abandoned, we're probably looking at the best place on the Ring for egress," I said.

"Agreed. Initiating landing sequence," Tayler radioed. "I'm setting us down near the machine."

Retro rockets on my JJ fired and once more, G forces pressed my body against the resistor suit. My stomach lurched as we began our descent.

Below us, the three-kilometer-wide machine drew closer. The machine itself was a marvel. An array of silos lined the leading edge like hoppers in a brewery. Spinning disks lined its front, ready to carve up the rock ahead of it, while pipes directed the slag to the north and south of the Ring. On the machine's roof, structural components rested on conveyors. Robotic riveters and welding arms were ready to secure each new component to the Ring.

"The machine was nuclear powered, but I don't see any radiators," I said. "Where do they get rid of the excess heat?"

"They pump it inside," Kevin answered. "Unlike Europa, Callisto doesn't have much tidal heating. There's no reason to vent the energy back into space."

I kept my eye on the machine as we landed half a kilometer to the south. Jupiter hung in the night sky like a Chinese lantern, giving everything a peculiar brownish- yellow glow.

We landed in the slag, the gravely leftovers from the dredging process of the Santa Claus Machine used to construct the Ring.

The augmented view of my surroundings disappeared, and for the first time in months I saw the sphere that had been my home. The suit released its magnetic hold on my body and I peeled it off, piece by piece.

Under gravity now, I moved freely in the sphere, wearing nothing but the fancy underwear. I ripped out the I.V. like a mother pulling off a Band-Aid. "Goddammit!" I yelled. The IV was more than a needle and had extended deep into my arm.

I took a deep breath and removed the catheters, expecting an equal amount of pain. Fortunately, because of our diet, we didn't make much number two during the trip. The special fluid we drank was designed to keep our stomach and digestive tracts in optimal condition. The catheter removal was less painful than I'd expected.

The sphere contained a storage unit where I found a NASA jumpsuit. A larger compartment held my spacesuit.

I put on the jumpsuit and then donned the spacesuit. With a systems check complete, I ordered the computer to decompress the cabin and open the door.

The seven others were already outside. Avro and Singer worked together, placing supplies on each other's backs. Rifles and rope hung off their oxygen tanks. They looked off balance but didn't fall, their suit's gyroscopes keeping them level.

Kevin joined me, and helped open a side panel on my ship. I retrieved my rifle, and Kevin placed it on my back like He-Man's sword.

"Anyone need any help?" I said, as Kevin and I joined the rest of the crew.

"Nice of you to join us," Serene said.

"Nice to be here. It's been a while," I joked.

"Come with me," Serene said. "We'll clear a path for the boys with the gear. Here, put this on." She handed me a harness. "For rappelling." She tossed one to Kevin as well.

"Thanks," I said, grabbing the harness and pulling it to my waist. We'd trained for this. If we had to break into the Ring, we knew it would probably be from above.

I took a few steps. The ground didn't quite feel right. It was almost too smooth.

"What's with the ground?" I asked.

"Solar polymer," Kevin answered. "Spray on solar panels, to supplement the nuclear reactors."

"Why didn't we use this on Mars again?" The question was rhetorical. This stuff only worked in a vacuum.

Avro, Singer, and Nash carried the bulk of the supplies, mostly extra guns and ammo. Amelia and Tayler carried packs filled with rations. Serene and I carried the rope. Kevin carried only a tablet, which he stared at, not paying attention to where he was going. Maybe the others thought him rude, but I could see his wheels turning. He was probably working on a plan that would save us a lot of hassle.

"That thing can move at what, two-and-a-half centimeters every second?' Avro said as we got closer.

"I'd say that's about right," Tayler replied.

"Two hundred and nineteen-thousand, four hundred and

fifty-six centimeters per day," Kevin said, glancing up from his tablet. "Just over one-point-six kilometers."

"So, five hundred and sixty-three kilometers per year," I said. "Impressive." I climbed over a large rock as we neared the constructor. Turning around, I offered a hand to Serene, which she grabbed, and I hoisted her up. This was the first time I'd held her real hand. It gave me a familiar sensation, like my first kiss in the sixth grade. Or holding hands with Marie on our first date. *Nice!*

Stairs zigzagged up the side of the machine, steps covered in dust.

Commander Tayler took the first step. He stood on the machine and gazed up as if looking for trouble. He then nodded for the rest of us to follow. We climbed past several landings where one would expect to find hatches leading into the body of the machine. There were none.

Jamaal Nash turned to me and said, "Does this thing have a door?"

"I don't know," I said. "It might only be accessible from the inside."

"Why don't we blow a hole in the side and check it out?"

"Oh no," I said. "We're not decompressing the machine to feed your curiosity."

We continued up until we reached the roof. Climbing this high with hundreds of pounds of gear would have been near impossible in Earth's gravity. But on tiny Callisto, the assent was a breeze.

Gangplanks on the roof provided an overhead view of the construction project. Gaps in the Ring exposed the surface below, where panels had yet to be attached.

"That's our way in," Tayler said, pointing at the gap. "Follow me." He climbed over the edge of the gangplank, stepping down amongst the conveyors.

I hopped down next, offering a hand to Serene. She grabbed it, and practically leapt into my arms as I lowered her onto the panel.

We laid out eight coils of rope, letting the ends dangles over the edge. We fastened the ropes to the base of several robotic welding arms and prepared for our descent.

"Let's do this," Nash said, his back facing the gap. He set his boots on the ledge and leaned back, transferring his weight to the rope. A moment later, he'd slipped over the edge. The rest of us followed his lead, sliding down to the surface.

In front of us, a barrier stretched from the floor to the ceiling. Steel girders connected the wall to the machine and a poly-fiber gasket contoured perfectly to the ground.

I took my first steps inside the Ring and looked at the others. "We're still inside the Universal Constructor," I said.

Amelia walked up to the wall, placing her hands on the giant gasket. "What is it?"

"It's the pressure barrier, the vinyl polymer that prevents decompression," Kevin said.

"Which means there's atmosphere on the other side," Luke Singer said.

"Yup," I agreed, looking around.

Jamaal Nash walked over and hammered his fist on the barrier. "So, you're saying if we blow a hole in it, it'll immediately stretch to fill the space."

"What is it with you and blowing holes?" Serene said.

"Just sayin' ..." Jemaal replied.

I looked around. A ladder rose over the rubber seal where the barrier met the wall. Tayler saw it, too. "Airlock, over here," he said.

We ascended to the airlock. I half expected to find it locked, but, as Tayler twisted the wheel, the hatch popped open. The airlock was barely large enough for all our gear. We squeezed in, and I closed the hatch behind us.

"Clear?" Tayler said.

"Clear," I replied. Tayler cranked a valve and atmosphere filled the room. A red light on the ceiling turned green as the airflow stopped. Avro spun the front hatch and it opened.

Avro stepped into the Ring, and the rest of us followed, to gather on a metallic platform connected to the pressure barrier. There was light on the other side, but not much.

Out to the horizon, and in every direction, stretched a brown nothingness of desert.

Upon first observation, the Ring appeared empty.

25

“Keep your helmets on,” Avro said, monitoring his augmented displays. I checked my screen. The air inside the Ring was breathable, but it swirled with a grimy fog, like dust behind a Mac-truck on a dry summer’s day.

“What’s causing the wind?” Amelia said.

“Temperature and pressure differentials,” I said. “The better question is, why is there wind?”

We stepped out of the airlock onto a metallic deck, as a waterfall of wet black dirt fell from the receding wall. A gable shielded us from the stream.

“Synthetic topsoil,” Kevin said, grasping a glove full of falling soil then tossing it out into the Ring. “The wind probably spreads it around. The extreme weather prevents vegetation from growing up near the barrier when the machine is not in use.”

“How do you know this?” Amelia said.

“Asimov,” Kevin replied.

We stepped down onto freshly laid dirt; the wind pelting our suits with dust like a heavy rain. I knelt down, also grabbing a handful of soil in my glove. It was peppered with tiny seeds.

We trudged further into the Ring, shaking synthetic topsoil from our suits, and monitoring the air quality readings. As we got father from the barrier, the wind began to calm.

In the distance, liquid water poured from the Ring's roof in heavy sheets. The water flowed into all the low-lying areas, creating a bog. Mist billowed above the water's surface, and a fog bank rolled in our direction, consuming us. Condensation formed on our helmets, and I found myself constantly wiping it away with my glove.

As we pressed on through the fog, I noticed the atmospheric rating on my HUD changing from yellow to green. The water had filtered the air.

I removed my helmet, and took a breath. The limestone aroma was like freshly mudded drywall. The air was so humid that droplets formed on my nose and began dripping into my suit. The fog cleared, and a lake shimmered in the distance, reflecting a grey, artificial sky. We definitely wouldn't be lacking water.

"I bet that half the power of that machine is dedicated to melting ice," I said.

"And electrolyzing the water to produce oxygen," Kevin added.

"What do they do with the hydrogen?" I said.

"Make complex carbon chains," Kevin said. "The carbon comes from the limestone."

"Organic compounds?" Amelia said.

"Sort of," Kevin replied. "They'd make fertilizer and printer filament first. The organic stuff takes care of itself."

I looked around, thinking deeply about what had been constructed here. The Ring, as a concept, was beginning to make a lot more sense. At first, the idea had seemed so grand, I'd thought it impossible. But now I realized something. The scale of the endeavor *allowed* for simplicity. *The Ring was its own ecosystem*, something that would have been impossible in a small habitat like the Martian colony.

As we walked further east, we began to transition to a living world, and the roof became blue. A holographic sun appeared, warming my face, and grass began popping up under our feet. Another kilometer or two, and shrubs and small trees grew past our knees. The further we went, the taller the trees got, until they were fully grown.

We walked until the holographic sun began to set at the horizon to the south. A river traced an S through the landscape while a dense forest fully covered the land. The trees were maple, but taller than any I'd ever seen. It must have been the low gravity. Down by the water, deer gathered on the shore, and I swore I saw a fish jump.

We left our spacesuits in a pile by the northern wall, and I walked with Serene down to the water. It was the first time we'd actually been alone together in the real world. Somehow, being with her in reality felt different. On the spacecraft, we had a joke: "What happens in VR stays in VR." I suddenly felt guilty, and thought of Marie. *It's like being unfaithful to her.* But I let the guilt pass, convincing myself that it served no purpose.

A stream trickled over rocks before flowing into the lake. I squatted down, cupping the water in my palm, and bringing it to my mouth. The water was fresh and freezing cold. It was probably the freshest water I'd ever had in years.

Dark clouds moved in waves along the sky, and the temperature dropped several degrees. "I think it's going to rain," I said.

Serene adjusted the rifle slung across her back and turned to look. "We should set up camp."

Moments later, it began to pour. It had been years since I'd felt rain. I closed my eyes and tilted my chin towards the sky, letting the drops stream down my face.

"You're awfully sentimental," Serene said, and began trudging back towards the trees.

"And *you've* been on Earth within the past two years," I said.

"I guess I'd miss the rain, too," she replied. "Help me make a shelter, would you?"

The rain lasted for an hour. We waited it out under a tarp draped over a branch. When the clouds departed, the last traces of daylight had vanished, and the stars, or the simulated projection of stars, appeared in the sky.

The temperature dropped to about twelve Celsius. Avro and Nash gathered wood and started a fire. Luke Singer had somehow caught a wild pig, which we cooked over a makeshift spit.

I helped Serene build a lean-to using a Mylar thermal coving. There were emergency blankets in our packs, and we made a bed. We slept side by side, experiencing real human contact; the first in months.

All I did was hold her hand; for now, that was enough. But my mind was accepting what my heart was fighting. Serene was now my girlfriend.

⚡

We awoke and ate a breakfast of apples and oatmeal rations. Commander Tayler discovered a *Coffea arabica* plant and roasted the beans over the fire. He dropped the roasted beans into his thermos, making a crude cup of java.

We sat around the fire drinking our morning brew and planning for the day ahead.

Luke activated his holographic chess board, and both Kevin and I lost to him at clock chess. I blamed the clock; I like time to plan ahead. "You think before the game begins," Luke said more than once. "When you play, play on instinct."

Kevin still wore his AR visor that he'd used to protect his eyes while cutting wood.

"Okay Luke, one more game," Kevin said.

The two sat by the fire. Kevin made the first move, pawn to king four, then knight to bishop three on his second.

"The Ruy Lopez," Luke said, smiling with approval. The two exchanged moves in rapid succession until, about forty seconds into the game, Luke swiped his hand across the board, smashing the holographic pieces into nonexistence. "No freaking way!"

"Why think before the game," Kevin said, "when you can have something else do the thinking for you?" He lifted his AR glasses and winked.

"You cheated," Singer said.

"Call it a home field advantage," Kevin said. "Like forcing us to play speed chess."

"Speaking of home field advantage," Amelia said, "do you think anyone knows we're here?"

"I don't think so," I replied. "If they did, they'd have sent someone to meet us by now."

"So, what's next? We just walk until we run into someone?"

Tayler said, "The wind is from the east. From what I can tell, it always blows west."

"We're going to build a raft," I said.

Tayler nodded. "We've all got hatchets in our survival kits. I recommend we get to work."

"Hatchets? You've got to be kidding me," Kevin said. "How low-tech is that?"

"You've got a better idea?" Amelia said.

"Yeah, I do," Kevin said. "How about a saw? Like this one that I used to cut firewood." Kevin held up the saw, which I figured was just another cool NASA tool we'd brought along for the trip, but this one had a small gas tank on the side. All our tools were electric.

"Where the hell did you get that!" Jamaal Nash said.

"You know those pipes along the wall? I followed them. Turns out, one of them is for filament. Metallics *and* synthetics."

"You found a 3D printer," I said.

"While you guys were hunting for rodents last night, I followed the pipes," Kevin said.

"Why didn't you print us a frickin' tent?" Serene complained.

"Come on, it's like twenty Celsius. It looks like every few miles there's a printing station embedded in the wall."

"Anything else you'd like to tell us?" Serene said.

"The printing stations have a bathroom," Kevin said.

"Now that would have been nice to know," Amelia said.

"Well," Tayler said, "I guess we'll print our boat then."

"You have got to be kidding me," Kevin said. "One of those pipes contains methane."

"Well, Kevin, what do you suggest?" Avro said. "This brush is too thick to allow any vehicle, so we're pretty much limited to walking, or traveling by water."

"By my count, three of us are aerospace engineers, and we *all* work for NASA. We're on a moon with barely any gravity. We're got access to an industrial 3D printer, and unlimited fuel."

"What's your point, Kevin?" I said.

"For the sake of the gods, we're building something that flies."

Avro smiled, and slapped Kevin on the back. Luke looked up from a chess game he was playing against the computer, shrugged his approval, and then made his next move.

Serene didn't look impressed at all. "Print planes?" she said.

"It makes sense," I said. "In this gravity, we could strap wings to our arms and fly like birds."

"This isn't Titan, John," Kevin said. "The planes will require a fixed-wing design."

"I was kidding," I said.

"What about the element of surprise?" said Nash. "We don't want to lose the upper hand."

Avro looked at Kevin. "Can you make a silent engine?"

Kevin nodded.

Luke Singer closed his chess board, grabbed the shovel from his pack, and scooped sand onto the fire.

Tayler grabbed his gear, slung it over his back, and said, "Kevin, show us the printer."

The printer wasn't much more than a filament pool and a laser. It sat in a metallic barn like an aircraft hangar from the Third Gulf War.

A display in front of the pool glowed with a generic screen saver. I touched the display, and the interface appeared. I scrolled through a list of preloaded designs.

"There's not much here," I said as Kevin walked over to join me. "It's mostly farm equipment. Tractor, combine, tractor, tractor. There's even a section for barns; here are roof trusses and shingles."

"I got this," Kevin said, pushing me to the side. He stood in front of the display, took off his wristwatch, and placed it the consul.

"You always have something up your sleeve, don't you?" I said.

"Yes," Kevin said, "my watch."

"No one's ever designed an aircraft for gravity this light," I said. "Where do we start?"

"Think of it as cooking," Kevin said. "I have several aircraft in my database. I'm going to borrow components from different planes to create an aircraft idea for these conditions: one hundred kilopascals of barometric pressure, and gravity of 1.236 meters per second, squared."

Kevin's plan gave me a strange *deja vu.* I'd seen this somewhere before. Then I remembered reading the history of Lockheed's Skunk Works division, led by aviation design legend:

Clarence "Kelly" Johnson. Back in the 1950's, Skunk Works had a shoestring budget, yet was tasked with the impossible: designing an aircraft that could fly higher than Russia's missiles. With off-the-shelf parts, and the front end of an F-104, a small team of engineers created the U2.

I leaned in to get a closer look. Kevin scrolled through several gliders until he came to rest on a Schleicher.

Amelia leaned over my shoulder to see what he was doing. "A glider?" she said.

"For the wings," Kevin replied. "A high aspect ratio will reduce the stall speed, enabling the plane to fly as slow as ten KPH." Kevin may very well have read the same book. The U2 had wings like a glider, vastly improving drag characteristics at high altitudes and slow speeds.

Amelia looked confused. "Aspect ratio?"

"Long wings," I said.

With a flick of a finger, Kevin tossed the Schleicher to the left, saving it for later. Next, he scrolled though a list of training aircraft, stopping at a Javelin Jet trainer. "For the cockpit," he said. The Javelin cockpit sat the pilot and copilot one ahead of the other while maintaining an aerodynamic profile.

"Simplicity is essential here. We'll have no retractable landing gear, no trip tabs, or even flaps. Just ailerons, elevators and rudder. A go-cart would be more complicated."

"I like it," I said. "Reminds me of my Katana."

"A Katana, good thinking, John, we'll use that for the T-tail." After adding the last aircraft, he chose a silenced Mark V engine and brought his "ingredients" to the foreground. He selected each aircraft in turn, grabbing the required components and fastening them to the new model until he had constructed his aircraft.

Kevin pivoted the image, inspecting his design from every angle, and then made the final adjustments to ensure the design was airworthy. "What do you think, John?" he said.

“I like it,” I said. “What’s it called?”

“I’m calling it the Shakuna Vimana, a flying machine from Hindu Sanskrit texts.”

“I’m not sure I can pronounce that,” Amelia said.

“It’s common to state the manufacture’s name before the designation,” I said, “so how about we call it the Patel - Shakuna Vimana, or P-SV.”

Kevin nodded his approval. “Just one more thing,” he said, opening a color pallet on the display.

Using gestures in the air above the display, Kevin coated the commander’s plane in blue and yellow, like a Blue Angel. The second, Avro’s plane, he decorated in green and white, colors of the Mexican soccer team. The third, Nash and Singer’s plane, he painted in a black and gold checkerboard pattern, placing a knight’s silhouette on the tail. My plane, he decorated yellow and orange, like the Pelican I flew back on Mars.

“Much better,” he said.

“So much for keeping a low profile,” I said.

Kevin executed a command and the printer shot a laser into the filament pool. The filament solidified upon contact with the beam. Components rose from the filament, ejecting onto a staging platform. The computer printed the fuselage first, and Avro and I carried it the assembly area in front of the barn. Next came the wings. We set these in the slots on the sides of the fuselage. Finally came the rocket engines, and empennage.

In minutes, we were rolling the first P-SV off the line. After admiring Kevin’s blue camo paint job, I climbed into the cockpit and grabbed the stick. It was tacky at first, but as I spun the ailerons and rudder through their paces, the action smoothed out nicely.

The plane only had two instruments, an airspeed indicator and a fuel gauge. We didn’t need an altimeter since the ceiling was only about 300 feet, nor did we need any navigation aids since we’d simply be traveling due east.

"Fill me up, Kevin." I said. "I think this bird's ready to fly."

Tayler, Johnson, Nash, and Singer stayed behind to assemble the next aircraft as it came off the line. Avro, Kevin, and Amelia pushed the completed P-SV to a nearby pasture that would serve as our runway.

"Alright everyone, stand back," I said, flipping a switch that opened the valves on the fuel tanks. I pulled the ignition chord, as if starting a lawn mower. Magnets spun around a coil creating a charge. As the cord whipped back into place, a spark leapt into the compression chamber.

"Ignition!" Kevin yelled over the muted tornado exiting the nozzle. I reached for the throttle, pushing it forward. The plane slid on its landing skid and, at only fifteen knots, it wobbled into the air.

I punched in the throttle and the rocket plane leapt forward to 100 knots, and then 200. Throwing the stick to the right, I pulled the plane into a steep left bank and cut the throttle. The plane bled off speed in the turn, and I settled into a manageable ninety knots, fifty feet from the Ring's ceiling.

With only half the gravity of Mars, and 100 times the air density, flying on Callisto was a breeze. I cut the throttle to zero and glided over Avro, Amelia, and Kevin. I spun the craft around, and touched down like a duck landing in the water.

Opening the canopy, I said, "Everything checks out. I think we've got a winner."

In less than two hours, we had assembled all four aircraft. Nash and Singer gathered the spacesuits, hiding them under a camouflage tarp behind the printing station. The rest of us loaded our gear behind the rear seats. After a quick meal, we climbed into the cockpits.

Kevin sat behind me in the blue plane. Amelia and Avro were in the red plane. Commander Tayler and Serene took the black plane, while Luke Singer and Jamaal Nash took the green one. Our ear buds allowed us to communicate.

I led our little squadron, taking to the sky first. The planes functioned flawlessly, sailing smoothly through the air.

"Keep it below one hundred fifty knots." Commander Tayler said. "These babies burn through fuel like a kid drinking chocolate milk and I'd like to keep refueling stops to a minimum."

"How far should we go?" I said. We'd traveled perhaps ten miles already, and had passed two printing stations, which we assumed had access to the hydrogen pipes just like the one we'd started from.

"It's about four hundred miles until we reach the point where the original mining colony was," Commander Tayler said. "If there *are* people living in this Ring, that's probably where they'll be. Any sign of civilization, any sign at all, and we turn around, and land. You got that?"

"Yes, sir!" we responded.

We'd been flying for half the day when we found civilization.

I banked toward the center of the Ring where a river expanded into a lake. Something bobbed in the water. "It looks like we've got activity ahead," I said. "Turning back now."

"I count three boats," Tayler said. "Coming around."

We banked away, and kept low over the treetops on the northern side of the water. Halfway through my turn, I looked back. The boats had started their engines, wakes bursting into white existence at their sterns.

"No indication of hostility," Nash said. "But it does look like they've seen us."

"Affirmative, they see us," Tayler said. "Orville, fly a wide arc; we need to know if they are friendly. Avro, check for settlements, then regroup east of the lake for landing."

Nash and I broke formation, following the coast until the lake narrowed back to a river.

"They appear to be pleasure cruisers," I said. "Two pontoon boats, and a speed boat. No uniforms, civilian clothes. People of all ethnicities."

Avro flew to the south west. "We've got a dock here, and a path leading to a house. I see a barn and several horses," he said. "More houses, barns and stables."

"I'm guessing the mining corporation made a freeliving deal with the unions," Serene said. "Paradise in exchange for forgoing a return trip."

"What else could it be?" I said. "The Ring is probably a secret retirement playground just like the Presidio on Mars."

"Roger that," Tayler said. "Form up, and we'll set down in that pasture by the docks."

"Copy," I said.

I looked to where the boats were heading east, away from the dock. Suddenly, several of them beached themselves on the bank, like whales in low tide. People jumped from the boats, running into the woods as if looking for cover.

"Hey, guys?" I said. "Those people are scared of us."

"And did anyone notice those trucks?" Singer said.

"What trucks?" Avro said.

"Three o'clock, on that ridge," Singer said, spitting words at double speed. "And two more, six o'clock on the north side of the lake."

I looked right, and saw the trucks sported green camouflage, blending in with the foliage. My eyes focused to see flashes of light emanating from their direction.

A moment later, bullets whizzed by, striking the roof of the Ring. Artificial sky rained down, pelting my P-SV with shards of flexi glass.

"Evade, evade," Tayler said. "Low and fast."

I dove toward the water. The bullets followed, puncturing the lake at intervals. Cranking the control stick left banked the plane around a peninsula, granting me temporary cover from the nearest truck.

Tayler pulled up and banked right. The bullets followed him, tracing an arched pattern of destruction across the glass sky.

Avro banked left, hitting the throttle and accelerating back in the direction we came. "They sure are confident their bullets won't breach the habitat," he said.

"But sure-as-*hell,* they'll breach your fuel tanks," I said.

Singer and Nash headed south, and the truck-mounted machine guns followed them through the air. As the bullets tore the fibrous sky to shreds, large chunks of the holographic material pelted the cockpit like hail before a tornado.

"Kevin, next time you design a plane, add some fucking guns," Serene yelled over the radio.

"Calm down everyone," Tayler said. "Head west and regroup."

"I'm hit," Nash radioed. "Plane is still functional, but I've got wind in the cockpit."

I flew in for a closer look; the area behind the pilot's seat had taken several hits.

"Singer, you okay in there?" I radioed. No response.

"Luke?" Nash yelled from the front seat. "Can you hear me, buddy?"

Flying wingtip to wingtip now, I could see into the cockpit of the other plane. The body inside leaned forward against the harness. "Tayler, John here, we've got a man down, repeat man down."

Tayler ordered, "Head back ten miles and set down near the printer where we just refueled. Copy?"

"Copy," we all said in sequence.

I banked west and flew parallel to the wall at 200 knots. After about five minutes, I spotted the printing station and set the aircraft down.

On the ground, Commander Tayler jumped from his cockpit, reached into the cargo hold and pulled out his rifle. He slung it around his back and ran towards Nash's plane as he was touching down. Jamaal and Luke's aircraft came to an abrupt stop. Nash tried to open the canopy, but it didn't budge. Tayler pounded the canopy's hinges with the butt of his rifle. The fuselage released its grip on the transparent dome. Tayler cranked it off, flipping it to the ground on the other side of the plane.

Jamaal Nash sprang out and turned to his friend, then threw up.

Kevin and I were next to show up at Nash's plane. Tayler held up his palm as if telling us not to rush. Avro and Amelia landed nearby. They jumped from their cockpit and ran to meet us.

Bullet holes punctuated the cockpit; three of them in the fuselage, and two in the canopy. Blood leaked from the bottom two holes.

"Go print some more shovels," Commander Tayler said. "Luke Singer is dead."

My mind reeled in confusion; it held a primal belief that Luke would be right back as soon as the program ended, a side effect of living in VR. *This was not a VR battle. Luke is not going to rematerialize. Ever.*

26

Nash was pretty shaken up, but stood guard with Serene while we dug a grave. We buried the body on a rise south of the printing barn. An orchard of blossoming apple trees blocked the spot from view. There were only a few feet of topsoil, so we dug as deep as we could, and then brought over additional dirt and grass to make a pile.

When the hole was ready, Avro and I lifted the body from the plane, setting it in a fiber-plastic casket Kevin had printed. He had engraved the top of the casket:

Luke W Singer 2040-2074
RIP

Nash and Serene went up to the gravesite to pay their respects, before meeting us back at the barn.

Tayler looked at Nash. "Destroy the planes," he said. "We don't want them falling into the wrong hands."

Kevin was about to protest, but I held his arm and shook my head. He kept his mouth shut.

"Print us some camo," Tayler said to Kevin. "Civilian clothes as well in case we need to go undercover. Amelia, help him."

Kevin nodded and walked over to the station. Amelia followed.

"There's a recycolizer in the barn," Tayler added. "Ditch anything you don't want to carry. That includes your NASA jumpsuits. Wear the camo, but put the civilian clothes in your packs. We'll stay hidden until we know what we're up against. We move out in twenty minutes."

We left the barn, walking single file, looking like a platoon of rebels marching along a ridge. Each of us carried a rifle, ammo, and a backpack.

"What now, Commander?" Jamaal asked. "There's an army out there searching for us. I recommend we head back to Earth and return with reinforcements."

"No," Tayler said. "This is what we trained for. Hell, we expected this! We're pushing on."

I said, "This attack has H3's name written all over it."

"And if we catch him, then what? Drag him around with a rope around his neck?" Serene said. Her voice was choked with suppressed grief, and sounded enraged.

"We lock him in Singer's ship. Without VR," Tayler said.

Serene gave a hard jerk of her head in assent. It was a suitable prison; without VR, the JJ's cockpit was a prison cell. But solitary confinement was a mild punishment for his crimes.

We crept through heavy brush high along a ridge, confident that our camo kept us hidden. Faces looked drawn and taut. Boots thumped the ground.

Down in the valley, a gun boat wound up the river while soldiers on four wheelers rolled along the grasslands.

We had walked nearly three miles when Tayler spoke. We could tell he'd been processing the situation, and we knew better than to interrupt his thoughts.

"Listen up," Tayler said. "I know you're angry, but we can't let that distract us from the mission. Right now, we need answers; that's our priority."

"Those bastards killed Singer," Nash said. "*Revenge* is my priority."

"No, it's not. We need to know *why* they attacked us," Tayler said. "That's why we're going to take a prisoner." Tayler paused to tap his temple, and then said, "Visors."

We huddled in a grove while Commander Tayler opened the mission planning application. I tapped my temple and the augmentation interface appeared in front of me. Our visors had recorded topographical information during our flight, and the tactical computers used this information to create a map displayed on a virtual table.

Known enemy positions appeared in red. The computer projected their routes with orange arrows. The four-wheeling patrols seemed to be following stream beds through the brush.

"We'll set up the ambush here," Tayler said, pointed at a ridge. "Serene, use your sniper rifle to take out the vehicles. I'll light them up with tracers."

"We're going to ambush them with firecrackers?" Serene said.

"It'll keep them occupied," the commander said. "Avro, Nash, sweep in and pick up a prisoner. John, Kevin, head to the southern wall and set up some charges on that ridge."

The commander used gestures to rotate the map, focusing in on a hill near the center of the Ring. He tapped a rock overhang shaped like a machine gun bunker from Omaha beach, and labeled it, *HQ*. Tayler continued, "Once everyone has completed their task, fall back to this location."

The scenario the commander described played out in front of us. Animated versions of ourselves ran through the simulation in fast forward.

We watched as two friendlies, representing me and Kevin,

moved to the south, avoiding several enemy positions.

"Any questions?" Tayler asked.

"Why am I always the decoy?" Kevin said.

"Any serious questions?" Tayler asked

"No, sir," we said in unison.

Kevin and I set out, running as fast as we could with our packs. We trusted our visors' proximity alert system to notify us if we got close to the enemy. I double-tapped my wrist as I ran, bringing up a live feed from the others' visors. We settled in, and planted the decoy chargers. I'd keep an eye on Serene's feed, and detonate the charges to correspond to her shots.

A squad of six soldiers on ATV's drove single file in the narrow streambed, right where the mission software had projected. Kevin and I were clear of our decoys, and had begun to make our way back.

I kept a hand over the detonator, hitting the digital trigger as Serene took her shots. She hammered the rear most vehicles first, driving the ATV into the dirt by the force of the round. The soldiers flew over their four-wheelers, landing on the rocky stream bed. She hit the leader in the leg, the bullet massing through his thigh, and hitting the fuel tank. His vehicle caught fire, and four of the other soldiers picked themselves off the rocks and stumbled over, pulling the injured man away from the flames.

A lone solider held back, limping in pain toward the cover of a tree. He held his left wrist after having been thrown from his vehicle. He watched through the barrage of tracer fire as his comrades tended to their friends.

Tayler lit up the divide between the lone soldier and the others, while Avro and Nash swept in. Avro wrapped his rifle around the man's neck from behind, while Nash stripped him of his weapons. They dragged him backward into the woods, letting him collapse into unconsciousness.

The charges Kevin and I had set continued to go off at

intervals, making it appear that the sniper fire originated from the far hills. Despite being the first to complete our task, we had the farthest to go, and were last to arrive at the recon point.

Avro, Amelia, and Jamaal met us as we approached the outcropping, which hid a series of small caverns.

"The hostage?" I said.

"Serene and Tayler are watching him, waiting for him to wake up," Avro said.

Amelia stood with her arms crossed. "According to the soldier's patch, his name is Howard Steiner. His uniform says "Callisto Defense Force". But, besides his watch, which we smashed so they can't track him, there's no other tech on him."

"We checked his wrist, no RFID either," Avro added.

"What about his weapons?" Kevin asked.

"Basic stuff, printed rifle, no computers, no guided bullets, no visors," Nash said.

"All stuff that could have been made here, nothing from Earth," Kevin said.

"Anything else?" I asked.

"Yeah, kid looks like he's eighteen," Amelia said.

"Great, so we picked up a child," I said.

"Perfect age for a soldier," Nash said. "At eighteen you feel invincible *and* you follow orders. After twenty-five, your perspective changes. Following orders? Fuck that shit."

I said, "He would have been just a boy when he left Earth. Who would bring a kid into deep space? That's just cruel!"

"We'll find out more when he wakes up," Amelia said.

The cave entrance was covered in ferns. We pushed them aside as we made our way in. Gurgling water trickled down at our feet. The cave was empty except for our gear.

"Kevin, you're doing the talking," Nash said.

"You're going to let Kevin conduct an interrogation?" I said.

Kevin looked pissed. "Dude, way to get them to blame the brown guy."

Amelia clarified, "The commander wants to hide our numbers, so we're only sending in one person."

"Then why me?" Kevin said. "Is it my accent?"

"It has nothing to do with your accent," Amelia replied. "It's because you're the least intimidating."

"I can be intimidating!" Kevin said.

I looked him in the eye. "No, you can't."

"What do you want me to say?" Kevin asked.

"We need you to listen to him," Amelia said. "When he asks questions, answer yes or no, don't reveal more than you have to. Got it?"

"Yeah, sure, Mrs. Professional Hostage Negotiator."

Serene and Tayler came up behind us. "He's begun to stir," Serene said.

"Kevin, get in there," Tayler ordered.

Kevin's visor sent us a direct feed. We watched the interrogation overlaid in 3d above our cave's floor. Howard Steiner sat on the floor, propped against a rock. His arms were tied behind his back, and a rope led from his wrists to a root hanging from the cave's ceiling. When Kevin arrived, the prisoner stood, as if at attention, and walked forward to the end of his rope.

Kevin paced back and forth, approaching the prisoner and looking him in the eye, and then walking back to the mouth of the cave. Usually Kevin doesn't shut up. Seeing him silent *was* intimidating. At first Howard looked confused, but then he scowled. It was obvious the prisoner couldn't tolerate that he wasn't been asked any questions.

"You came from Mars," Howard said.

Kevin stopped pacing, and held his hands behind his back. *How the hell did he know that?* I thought.

"Yes," Kevin answered.

"Then, it's true. You're from the Communist Alliance."

Kevin unclasped his hands. "Hold please." Kevin left his hostage and met us in the adjacent cave.

"He just told me I was from the CA," Kevin whispered.

"We heard him, but what the hell?" Serene said. "The CA doesn't have space ships! Keep him talking. Don't respond to misinformation. That will just reinforce his beliefs."

Kevin walked back to the hostage, and stood like a solider at ease in front of him. Howard looked confused. He wanted answers, too.

"You're a commie," Steiner said.

"So you say," Kevin replied smoothly. "What do you know about us?"

"We know all about you," Howard said. "You've come to take Callisto, because the Martian habitats are too small, too old, and too shitty, so you've come to take our world, to make us live by your rules, or kill us."

"Who is your commanding officer?"

"I'm not going to tell you," Howard answered, and spat on the cave floor.

"Is it H3? I was always a fan of his. Nice guy in a murder-y psychopathic kind of way."

In the adjacent tunnel, I put a hand to my head, "Oh, Kevin," I said.

"Is this a joke?" Howard said.

"How did you get to Callisto?" Kevin asked, regaining his composure.

"In a space ship, you idiot; how did you get here?"

"Me too," Kevin said, sitting down on a rock. "It was a long trip, but the inflight entertainment wasn't bad."

An awkward silence passed. The boy appeared confused.

Apparently, Kevin wasn't the villain he expected. Howard struggled as if trying to escape, but then winced in pain as he put strain on his injured wrist. His scowl returned.

"Where were you from, on Earth?" Kevin asked.

"New York, not that it matters to you. I was an exchange student in Hong Kong. I bet you liked it when you nuked Earth, you communist scum bag. They say most of the weapons were yours."

"Ahh, hold one second," Kevin said, and returned to our cave.

"How did it feel when you killed those people on Mars?" Howard yelled as Kevin left.

When Howard was out of earshot, Kevin said, "So we nuked Earth, and we killed people on Mars? He called me a communist. What the hell, guys?"

"Shit," I said.

"What?" Kevin asked again.

"Don't you see what this means? This kid was kidnapped. Whoever brought him here wanted an army so they kidnapped a bunch of people from Earth under the guise of nuclear war."

"I'm going in there," Serene said.

"Me too," Amelia said.

"Wait," Tayler said. "Leave the camo."

Kevin, Amelia and Serene shrugged off their tunics. Underneath, Amelia and Serene wore blue T-shirts with the NASA emblem embroidered on the upper left. Kevin removed his tunic as well. He didn't have a NASA emblem; instead his shirt showed a squadron of B3 bombers flying in formation in front of a giant American flag.

"That'll do," Avro said. "Good luck."

The three of them stood in front of the hostage; it was dark in the cave, and their faces and T-shirts were barely visible.

"I'm Serene Johnson. Former US Marine. Nice to meet you."

"Amelia Shephard, former Lieutenant in the Multinational Defense Force."

"Kevin Patel. I work for NASA."

"We all work for NASA now," Amelia said, taking off her watch and using its light to illuminate the cave. "We were exploring the habitat when you attacked us."

Howard looked even more confused now. He stared at the women, maybe aware that they didn't have foreign accents, and weren't wearing the drab costumes of Alliance citizens.

"I don't understand," Howard said. "This must be a deception. You're trying to trick me. NASA doesn't use guns."

"Well, thanks to H3, we do now," Amelia said.

"You say you're American?" Howard asked.

"Actually, I'm Canadian," Serene said. "By birth anyway." She let her voice become soft, a change from her usual macho tone.

"I'm American," Amelia said.

"Okay then, how did you …" Howard paused, softening to the idea, "… survive?"

"Survive what?" Amelia said.

"The war, the apocalypse, everything!" Howard demanded.

"What war?" Kevin said. "There have been no wars."

"July twentieth, 2071—" Howard began.

"Impact day," Amelia said.

"Impact day?" Howard said. "What's impact day?" Howard shifted his gaze from Amelia to Serene, then back again. His look was one of innocent wonder, like a child face to face with a whale at an aquarium.

Amelia turned to Serene. "They were definitely kidnapped."

"Kidnapped?" Howard said. "We were rescued. I saw the nukes from the plane!"

Amelia asked, "Where were you on July twentieth, 2071, the

day the *CTS Bradbury* crashed into California?"

"In the air, somewhere over Asia."

Serene turned to Amelia. "The Chinese detonated several nuclear warheads shortly after the impact," she explained, "as a warning to the Communist Alliance. The Chinese were worried the Alliance would use the disaster to start an uprising."

Amelia sat down on a rock, looking up at Howard who still stood at the end of his lead. She picked up a stick, and began drawing on the cave floor. "There was a cartel in California which used the impact to come to power. In that sense, the Chinese were smart, as they may have avoided a war. Unfortunately, several transcontinental flights were caught in the crossfire."

"Except they weren't," Serene said. "You were on one of those flights."

"We landed in Tibet," Howard said. "That's where we boarded the spacecraft. But if there was no war," he continued, taking two steps back, "then Earth is …"

"A million people died that day," Amelia said.

"They told us Earth was uninhabitable."

"Who told you?" Serene said.

"The Doomsdayers."

"The Doomsdayers? Those conspiracy theorists back on Earth?" Amelia said.

"More like trillionaire executives preparing for the end of the world," Kevin explained. "Bat-shit crazy people with money. Rumor had it, H3 was a Doomsdayer, although he never stated this publicly."

Howard went to his knees. He was weeping, realizing that his life was built on a lie.

Commander Tayler stood in the entrance of the cave. "Untie him," he said.

Tayler looked back at Avro, Nash, and me. "C'mon in."

Avro lit a torch from his pack, fully illuminating the cave

Serene walked over to Howard. “May I?” she said. Howard turned, and presented his hands behind his back. Serene pulled a knife from its sheath, cutting the zip ties that bound his hands together.

They had wrapped his left wrist, which may have been broken, but his right looked fine. We exchanged introductions and shook hands. I could feel Howard’s hand tremble as we shook; he was probably still in shock.

“Take a seat everyone,” Tayler said. “Howard, tell us everything you know. Absolutely everything.”

27

Only a few months after its creation, the army was ready for combat. Several regiments had completed basic training. On Marie's morning runs, she'd pass platoons standing at attention outside their barracks. During the day, they'd march up and down the streets of their newly constructed base.

In Newport, the sound of gunfire from the range could be heard.

Several VR suits from the *Mount Everest* had been collected, and a Virtual Reality training school had been established to train pilots for spacecraft that had yet to be built.

H3 met with Commander Yamamoto, the highest-ranking officer in this army. The commander was just over a meter and a half tall, and would probably be less than this under Earth's gravity. He looked like a man who'd been religious about going to the gym in the past, but had cut back on his routine after taking on additional responsibilities.

They sat in H3's office aboard his yacht. Marie sat nearby working through logistical details pertaining to H3's position. Her side project however, was making sure to remind every politician, including H3, of her idea for a reconnaissance mission to Earth. It was still this that motivated her work for H3.

"Five thousand adults on this colony, none of whom are

aerospace engineers," H3 complained. "We need to defend the space around the colony, commander, not just inside it."

"We have our smartest men training in VR aboard the *Mount Everest*, sir, and a manufacturing facility has been constructed. What we lack are effective designs."

The two stared at each other.

Yamamoto said, "Sir, if we could use your cruiser as a template, perhaps we could replicate the design."

"No," H3 said. "No one touches my ship."

"Henry," Marie said, butting into a conversation well below her pay grade.

"Your ship made the trip all the way from Mars; at least let them take a look."

"Trust me, Marie, it won't help. My ship is a horrible template."

"I disagree, representative," Yamamoto replied. "You should listen to your assistant."

"I've been asking for a reconnaissance mission to Earth for months," Marie said. "If I had my way, I'd take his ship, and make the trip back to Earth tomorrow."

"Marie," H3 began with a scowl, "my spacecraft can only carry enough fuel for a one-way trip. Without a convoy, they'd never make it back. I'm not sending good men and women on a suicide mission just to please you."

There was a knock on the door, and a young lieutenant stepped into the room. He drew a quick salute which he held in place. Commander Yamamoto stood, returning the salute.

"Sir," the young man said. "Our patrol out on the eastern Ring has spotted eight spacecraft parked almost five hundred kilometers further to the east."

"What?" H3 said. "Where are they, specifically?"

"They're near the universal constructer, sir. The ships are

small, look like single pilot aircraft, and seem to be abandoned. We've followed all your protocols sir, and a radio jamming is in effect."

Yamamotto was speechless; he just stared at the young solider.

"There's more, sir," said the young man. "We've spotted several flying machines. *inside* the Ring."

"Oh my God!" Marie said, thinking of Branson and Lise.

"Well, take them out!" said Yamamotto.

"Yes, sir." The young man turned, walked out the door and began issuing commands into his watch.

"Shit," H3 said, banging his hand on the table. "They're here."

Marie studied him. It was the second time she'd seen him lose his temper. The first was when Malcom tried to send a signal.

"You must control your anger, representative," Yamamotto said. "The enemy can use that to their advantage.

"It's not anger, Commander, it's fear."

A siren sounded in Newport, and echoed off the water. Commander Yamamoto was just about to leave H3's office when he received a message.

H3 stood, and Marie joined him in front of his desk.

Yamamoto passed his hand over his wrist, transferring the message up onto the holovision.

"Captain," Yamamoto said to the screen.

The screen showed a military truck with a rear mounted 20mm cannon. The captain, a woman of about thirty, positioned herself in front of the camera.

"Commander Yamamoto," she said. "We took heavy fire from the south west. It was an ambush, sir."

"Was anyone killed?" Yamamoto said.

"No KIA sir, but one person is missing," replied the captain. "His name was Steiner, sir."

"No KIA? You took heavy fire and no one was killed?" the commander said.

"We had one person in critical, sir, six with minor injuries, and one soldier is missing. We believe Steiner may have been taken hostage."

"Keep me updated," said the commander.

"Yes sir," the captain said. "One more thing."

"Yes, what is it?" said the commander.

"Those aircraft we saw, sir, looked like they were printed here. The pilots destroyed them, sir, using high explosives on the hydrogen tanks. But we found something in the debris."

"Oh?"

"Fabric, sir. It was pretty bloody, didn't find a body, but we're pretty sure one of them is dead. The fabric had a patch on it. It said "NASA", sir."

"A souvenir from Mars most likely," H3 said. "The Alliance guys love their battle trophies. Thank you, captain."

H3 reached over to Yamamoto and tapped his wrist, ending the call.

Commander Yamamoto turned to face H3. "Sir, do you have something to say?"

"This is far worse than we thought," H3 said. "Far, far worse. You must exterminate this threat immediately!"

"No one was killed!" Marie said. "Maybe this means they don't want to kill anyone!"

"It means they're sloppy," H3 said.

"Representative," said the commander, "I'll send every man we have to the eastern edge of the settlement."

"That won't be enough," H3 said. "Torch the woods, burn it from the east to the west."

"Representative?" Yamamoto said.

"The colony will be fine, it's designed to survive a fire," H3 replied. "It is not designed to survive an occupation, Commander."

"Yes, sir," said the commander.

"And if you find your missing solider, hold him for interrogation but don't let him speak to anyone. These people have an amazing ability to corrupt young minds; they're probably trying to take us from the inside."

"I understand, sir."

"Carry on."

28

We sat on rocks around the cave, each of us facing Howard in anticipation. In the torch light, we looked like campers about to tell ghost stories. Kevin opened a ration pack and passed out bars of dried fruit. Howard talked for over an hour and he turned out to be quite a story teller. He told us about the Hives and the four spaceships that left Earth, and the three that made it to Callisto. It was almost unbelievable. Except, here they were, thousands of people living their lives in a tube. The baby boom surprised me the most; almost 1,000 babies had been born since they arrived, and many more were on the way.

"That's it really," Howard said, bringing his story to a close. "I joined the Callisto Defense Force, and we've been patrolling ever since. When you showed up, we thought it meant war."

"How many CDF soldiers are there?" Serene asked.

"About five hundred," Howard replied.

"Where did they get the officers?" Avro asked.

"They were chosen personally by H3, and all of them have previous military experience. But mostly, they were NCOs on Earth."

"It sounds like H3 wanted an army that wouldn't question orders," I said.

"I never really thought about it," Howard said.

"What made you join the CDF?" I asked.

"There's not a lot of opportunity for adventure here on Callisto. When we heard they were forming an army to protect the citizens, all my friends and I signed up."

"Are there many other Americans in the colony?" I asked.

"A few hundred maybe, but we are all pretty spread out now. Some were on flights like me; others were on the cruise ship that escaped the tsunami. I was fourteen when we left Earth, and don't think about nationalities anymore."

"I lost my wife and son in the impact," I said.

"I'm sorry," Howard said. "But if what you're saying is true, my parents, and my siblings, are still alive. Can I contact them?"

Serene tapped her watch. "Our comm equipment won't penetrate the walls of this colony, sorry."

"Now that you're here, you can tell everyone the truth, right?" Howard said, "Maybe some of us can go home?"

"I don't think they're going to let us get the story out," Commander Tayler said. "H3 is a wanted man and the Doomsdayers obviously don't want anyone finding out the truth."

"They can't keep this a secret forever; they must have known that," Howard said.

"They do know that," I said. "That explains why they wanted a birth boom. Anyone born here would have serious trouble adapting to Earth's gravity."

"So, anyone with family here is stuck here no matter what?" Howard said.

"Basically, yeah," I agreed.

"What would they do if they found you?" Howard asked.

"They'll kill us," Serene said. "And since you know the truth, in all probability, they'll kill you, too."

"I'll deny it, "Howard said.

"They'll know you've talked to us," Tayler said. "I'm sorry

we brought you into this."

Our proximity alerts began to chirp. "We've got company," I said, and pulled on my visor. Half the screen showed a swirling stripe of red.

"Tactical update," Avro said.

Nash grabbed a drone from his belt and tossed it out of the cave. Four little propellers popped out of the side and the drone flew a kilometer loop around our position before flying back into the cave, and Jamaal's palm.

The new data fed our visors, but something was wrong. Something was clouding the drone's infrared sensor.

"What's going on? Is your visor acting up, too?" Amelia said.

Tayler stood up, and grabbed his gun. "Howard, what happens if there's a forest fire inside the Ring?" he said, turning and walking over to the entrance of the cave.

"It rains often enough," Howard answered. "It gets pretty wet in here, and I assume that would put it out, eventually?"

"Eventually," Tayler said. "Eventually is not soon enough. They're burning the whole damned forest."

Smoke billowed from the treetops as flames began climbing the hills. The smoke rose high in the sky, curling as it hit the ceiling. In the distance, steam poured from the Ring's wall, creating the clouds that would eventually douse the flames, but it would be too late.

"We gotta go," Tayler said. "Now!" Amelia helped Howard to his feet while the rest of us ran to the adjacent tunnel. Serene grabbed her pack, sniper rifle attached to its side. I helped her sling it onto her shoulders.

"We're out of time," Amelia yelled into our cave as smoke began billowing into the entrance. I grabbed my pack, weapons fixed to the sides. I slung it on as I ran, leaving my tunic and ration packs behind. Avro and the other men followed me out, weapons in hand. Nash poured water from a canteen over his clothes as he ran. I pulled my canteen from my side, and did the same.

The smoke thickened and I lifted my dampened shirt to cover my face. We followed the stream that emanated from the caves, our feet splashing as we hustled down the hill. But the stream was exactly where they'd expect us to go, a bottleneck.

Tayler motioned for us to follow him, to flank whoever waited for us. Trees and brush burnt around us, and I coughed. The air was thick with ash that flew around like snow in a blizzard. We jumped high over a jumbled wall of burning brush, a simple feat in Callisto's gravity, but a stunt that risked putting us out in the open. When we landed, Kevin's pants were on fire; he opened his canteen, pouring water on the flames.

A dozen soldiers waited by the shore, pointing their guns at the stream. A beach craft rested in knee deep water. They didn't see us. We slid below a felled log further up the beach.

"Please don't kill them, they're innocent!" Howard whispered loudly.

"The hell they are," said Serene.

"They killed my best friend," Jamaal added.

"I … I didn't know they killed anyone," Howard said.

"Nash, shut it," Tayler whispered. He took a moment to load a clip into his rifle. "We need to find a way around."

"Not going to happen," Serene said. "Hold him back!" she yelled. Nash grabbed Howard by the arms,

"No!" our hostage yelled, and soldiers on the beach began to turn in our direction.

In a fluid motion, Serene unslung her rifle from her pack, flipped the weapon into automatic fire mode, and held the trigger.

"Serene, no!" I yelled. Tyler slapped a hand over my mouth.

Guided bullets flew toward the waiting soldiers, each one

finding a home.

The platoon dropped. We hopped over the log, and ran toward the fallen soldiers.

"Gather their weapons," Tayler said. "Throw them in the boat. Their watches will report them as deceased in thirty seconds, so get them off and place them on your own arms."

"Each of you, grab a tunic," he continued. "The more we can do to confuse them, the better."

We pulled off their blue uniforms.

Nash was still dragging Howard along. The boy was distraught; this was probably the first time he'd seen death.

The commander walked over to him, slapping a watch on his wrist. "Howard, get on the radio," the commander said. "Tell them you've been ambushed and direct them away from this location."

"I … I can't," Howard said. "You killed them! You killed all of them!"

"Kill or be killed, Howard. You're on our side now. We need you. We need your help to free this colony."

Howard nodded, bringing his wrist to his mouth. I half expected him to give us away. "This is Private Steiner. We've been ambushed, enemy moving north east, I repeat enemy moving north east." He released the transmitter.

"Good job," Tayler said then unstrapped the watch from his wrist.

I pulled a bloodied Callisto Defense Force tunic over my own, and helped the others do the same.

"I assume you want to head to the capital," Howard said.

"Is that where H3 is?" Commander Tayler said.

"Probably," Howard said.

"How far is it?" Tayler asked.

"About twenty klicks," Howard replied. "The river leads directly to Newport."

"Then I suggest we steal this boat."

The beach craft was grey, the color of metallic filament used by the printers. The barge opened with a ramp at the rear that doubled as the stern. The ramp lay in ten centimeters of water.

At our height above the water, the horizon was less than three kilometers distant, and it was only a matter of minutes before several other boats came into view. Fortunately, they moved into the shore, far from our location, lowering their ramps as platoons of troops drove off on ATVs.

Avro entered the wheelhouse and took the helm. Inside were minimal controls; the barge was printed from a primitive design. The boat purred to life as Avro adjusted the throttle. Nash and I pulled up the ramp. Tayler collected the watches and tossed them overboard. This would draw the enemy toward this location after we'd departed.

We motored along the shore, trying not to look suspicious. Smoke billowed along the Ring's ceiling, mixing with the rain clouds. The sky went dark as night, and it began to pour. Steam rose from the hills as the rain began to subdue the fires. Sooty rain pelted us with grey droplets that stained our clothes.

The rain turned to a downpour and the surface of the river swirled with mist, visibility dropping to a few hundred feet. Under the blanket of fog, Avro gunned the hydrogen-powered boat toward the capital.

"Keep an eye out for those guns on the shore," the commander said." It's only a matter of time before they realize we've got their boat."

"I don't like this, sir. We're an easy target for their artillery," Nash said.

"They don't have artillery," Howard said. "The sky is too

low."

Avro turned to Howard. "Then what do you have?"

"Guns, lots of guns," Howard replied. "A few hover-platforms, no airplanes though. Those jets you built were impressive."

“Hover-platforms?” Kevin said.

“Levitating platforms for repairing the Ring,” Howard replied. “They’re jet powered, and the only thing on this rock that flies.”

“And communication equipment?" Amelia said.

"Radio app in our watches, a few directional beacons, and a device for signal triangulation. You don't need navigation aids when you live in a tube."

After five miles, we entered a more populated part of the Ring. On our right, the town rose from the hillside, with pastures, and Cape Cod-style houses. Behind the houses were more barns and pastures where black, brown, and multi-colored horses pranced. The rain had stopped, and the holographic sun illuminated the lush grasslands.

"Where are we?" I asked. "It’s beautiful."

"The town is called Clydesdale. It's where the animal lovers live," Howard explained.

“Animal lovers?” I said.

“People formed towns based on their interests and values. There are Japanese tea garden villages to the south; Callisto has quite a bit of diversity.”

A road ran along the riverbank, and several people sped by on bicycles. A pleasure craft zipped by, heading for shore. Avro waved to the occupants in the other boat, and they waved back.

"Any ideas where H3 is hiding?" Commander Tayler asked.

"He works from a boat docked in the harbor," Howard said.

"Well, that's convenient."

"Hey guys, is this barge bulletproof?" Kevin asked.

"Probably," Amelia answered. "The sides are anyway." She pounded the barge's wall with her fist. "Why?"

"Because I think they've found us."

On the north side of the river, a military green truck matched our speed as it rumbled down the road. From the truck's bed, a twenty-millimeter cannon erupted to life.

"Get down!" Nash yelled. We ducked under the gunwales as bullets whizzed over our heads. Indents formed in the hull where bullets impacted, but the armor held.

Avro ducked as bullets shattered the wheelhouse, covering him with glass. He knelt next to the helm, peeking out the wheelhouse window to see where we were going. His face was bloody from a cut on his nose and cheek. Amelia crawled forward, grabbing a first aid kit. She wiped off the blood and sprayed his wounds with liquid skin.

We moved through the water at about twenty-five knots, a speed the truck on shore could match. "Got any ideas?" Avro said, turning back toward us.

"Hug the southern shoreline," Serene suggested. "Doesn't look like there's a road on that side of the river."

"Roger that." Avro yanked the wheel to port, and we banked left as bullets continued to impact the side of the boat.

"We're about to have company," Nash said. "Bogeys, two o'clock."

I peeked over the side. The gun fire from shore ceased as two speed boats approached at over fifty knots. I counted the soldiers. "Looks like five men in each boat."

"Your friends carry grenades?" Nash shouted at Howard.

"Hand grenades," Howard answered.

"Launchers?"

"No, just hand grenades," Howard said.

"Good. 'Cause we do." Nash flipped a switch on his rifle and loaded a grenade into the barrel.

Tayler said, "Slow down, let them get close, but not too close."

Avro banked the barge around, cutting the throttle. Our pursuers slowed to match our speed as they approached the barge. The soldiers stood behind their windshield, grenades in hand, ready to throw when they were in range.

"Anytime now, Nash," Serene said.

Jamaal stood up, firing at the closest boat. The grenade impacted on top the boat's bow, exploding in a flash, but the boat maintained its pursuit. He launched another, and the boat banked left. The grenade hit the water and a jet of spray shot into the air.

"Give me that," Serene yelled, taking the gun from Nash. She lined up a shot, aiming ahead of the boat. The projectile landed in the cockpit. Soldiers dove from the boat as the grenade hit the deck, bouncing once before exploding. Hydrogen tanks ruptured, and the boat disintegrated as the atmosphere rippled with fire and heat.

The second boat retreated, following out of range of our launcher. Avro banked around, and began heading west again.

Newport's skyline rose above the not so distant horizon. Cobblestone streets wound through European-style buildings while birch trees lined the thoroughfares. Twenty CDF soldiers hovered in the air, levitating the primitive hydrogen platforms Howard had mentioned. This was their last line of defense.

Nash peeked over the gunwale, and a barrage of fire rained toward the platforms. Avro tried to keep his distance; any closer, and we'd be sitting ducks.

Buildings rose along the shore, and people watched from inside the windows. "If we start firing, the whole town will think we're terrorists!" I said.

"If we don't, we're all dead," Nash retorted.

The commander looked at Serene. "How's your aim?" he asked.

"Do you have to ask?" She propped her sniper rifle on the bow.

"The townspeople are watching. Keep that in mind," the commander ordered.

Serene rested her back against the forward bulkhead, and set her rifle on the hull facing the opposite direction. It was the most awkward and ingenious firing position I'd ever seen. She took careful aim, using the sight on her visor to hone in on the targets while keeping her head safely below the gunwale. Our barge bobbed up and down, but she compensated with her breathing, letting the rifle rise and fall with the rhythm of the boat. She fired. The bullet ripped through one of the platforms, sending the craft spiraling out of control. The solider jumped, diving into the water below. She took aim at the next platform, but the bullet missed it, hitting the soldier in the calf. His leg was blown clear off, and his body spiraled off the platform. Serene didn't even pause.

“Dammit, Serene,” the commander said.

She moved onto her next target, and then the next, until all the sky above us was clear.

Several soldiers trod water as Avro raced through the defensive perimeter, careful to avoid those in the drink. Howard looked distraught. He must have felt how impossible this situation seemed. Even if we convinced the colony we were from NASA, we still had blood on our hands.

"What now?" Amelia asked of no one in particular.

“We need to get the message out. Proving we're from NASA is our best defense,” I said.

“Agreed,” Tayler said, “but we need to reach everyone at once. Otherwise, H3 might kill anyone who holds the information.”

“H3 broadcasts from his boat,” Howard said. “I'm sure he's got tech in there that no one else has. Holovisions and other stuff from his spaceship."

"Broadcast?" Kevin said. "To what network?"

"Our watches are linked through a social network called Cal-Net," Howard said. "Anyone with a connection has the ability to post messages to the entire colony."

"If H3's got the hardware, I'm willing to bet we can log in."

"Is that H3's boat?" Avro asked, pointing to a large black mass of hull and tempered blue glass. The yacht was docked at the end of a wharf extending several hundred feet from shore.

"Do you really have to ask?" Howard said. "I bet H3 was as eccentric on Mars as he is on Callisto."

"Oh, you have no idea," Amelia said. We approached the boat cautiously. It was high off the water, and anyone on board would have a perfect line of sight into our barge.

It was clear where we were headed, and a platoon of soldiers began running toward the wharf, taking cover behind wooden stanchions. Serene and Avro fired warning shots at the dock to slow their progress.

Nash pulled the surveillance drone from his pocket and tossed it into the air. I tapped my visor, adding the drone's camera feed to my field of view.

Howard said, "I haven't seen one of those since I was a kid!" He craned his neck, following the drone through the air. "The people that founded this place were against anything that resembled drones or AI."

Nash shrugged off Howard's comment, piloting the drone up and above the yacht.

"Clear on deck. Infrared shows no bodies in the cabin," Nash reported.

Through the drone's eye, we could see deep into the boat. The cabin was open concept and surrounded a flexi-glass enclosure. A staircase led to the lower levels, a replica of the grand hall from the

Titanic.

"He's probably retreated to the capitol building," Howard said.

"Move in," Tayler said, "but watch that fire from the shore."

Avro steered parallel to the yacht, making sure to drag our leading edge along the yacht's hull, drawing a sizable scratch in the paint.

Nash jumped, grabbing the bow railing, and climbing onto the deck. Amelia tossed him a line which he secured to the rail. He found an emergency rope ladder and threw it over the side. Serene and Amelia climbed up first, followed by Kevin and Avro. Tayler and I tossed our remaining ammo packs onto the deck, before climbing up ourselves.

It was clear the entire boat was printed, but tastefully so. Keeping low to avoid the bullets from shore, we ran across the fiber-plastic mahogany deck. The cabin's atrium rose above the deck, providing cover from any bullets coming from shore.

The cockpit was locked. Avro reached into his pack, grabbed a charge, and blew the door. He used the butt of his rifle to smash his way in.

Kevin placed a countermeasure detection system on the glass. "Clear," he yelled.

"Clear," Nash yelled, as he went inside. The drone followed him in, and he piloted it through the interior. Inside, there was a large bedroom in the bow, but most of the ship held offices.

It was obvious which one was H3's. A tall wingback chair sat behind a large mahogany desk. Holovisions on the walls displayed hypnotic images of morphing art, including Salvador Dalí's painting of melting clocks.

Nash summoned his drone back. It landed in his palm and folded into a sunglass case shape. "I'll stand guard," he said.

"I'll join you," Serene said.

"Avro, cut the dock lines," Tayler ordered. "I'll use the barge's engines to pull us out."

Bullets clinked off the boat's hull and I wondered how much that would piss off H3; probably not much. If he won this battle, he'd probably just print a new one.

Kevin, Amelia, Howard, and I entered H3's office.

"Here," Kevin said, pointing at a large display on the wall in front of H3's desk. He swiped away the screen saver and accessed the menus. The screen went blank. "Identity confirmation required," came a computer-generated voice.

Kevin clasped his hands, shutting off the system.

"I'm completing a reset," Kevin said. "When the system comes online again, we'll generate a new social media profile."

"Let me know if you need me to come up with a good password," Amelia joked.

"Biometrics," Kevin said. "His computer uses Turing Intelligence to confirm the user's identity. It knows you by your look, your mannerisms, the same way your mother knows you're her daughter."

"Please don't bring my mother into this conversation," Amelia said.

"This should do the trick," he said, removing his visor and reactivating the display. The screen showed an augmented reflection of H3's office, including Kevin and me.

"We're in," Kevin said.

"We're in?" I said, waving at our image on the screen and watching my projection wave back.

"Welcome, Kevin," said the holovision. The Picassos on the walls disappeared and security camera footage displayed on the screens, letting us see the battle raging outside.

Howard pointed at the screen. "Cal-Net Social," he said, and a stream of videos and images scrolled along the side of the screen.

Most of the news was of us: images of our boat as seen from shore, pictures of the airplanes. I read the captions to myself:

Who are these people?

Is this really the CA?

Clydesdale under attack!

"Get that message out before we sink!" Nash yelled.

The screen to my left showed the starboard side of the vessel. Two barges filled with soldiers approached. On our deck, Serene loaded a grenade into her launcher and fired at the first boat. The grenade landed inside the barge and exploded, obliterating its occupants. She reloaded and hit the second boat although it swerved to avoid the projectile. She reloaded and fired again. The grenade hit the side of the boat, and exploded centimeters below the surface. The barge lifted out of the water, flipping onto its side.

"We're ready, when you are," Kevin said.

"Hit it," I said.

Kevin used a two-fingered gesture to start the broadcast. A red circle appeared in the bottom corner of the screen.

"Are we live?" Amelia said. Kevin nodded the affirmative.

I said, "If anyone can hear us, we are not your enemy. We're representatives from NASA. My name is John Orville, this is Amelia, and Kevin. Please, stop firing on this boat. I repeat, we are representatives from NASA."

"Please listen," Amelia said. "There was no Doomsday, and you are being lied to. Please, if you can hear this, stop firing on H3's boat."

"I'm from the CDF," Howard said. "I can confirm their story. These people are not from the Alliance."

"They're still shooting at us!" Nash yelled from up top.

Nash unloaded a clip onto the dock; several of the soldiers were hit, and three of them tumbled into the water.

"They don't mean to hurt anyone," Howard yelled at the

screen, tears streaming from his eyes. He glanced to the left as several CDF soldiers on the dock took bullets to their chests. “They’re fighting to defend themselves!”

“Keep talking,” Amelia suggested. “Give it a few minutes for word to spread.”

“Please,” I begged, “please tell the soldiers to stand down.”

Howard said to the screen, “It’s self-defense, I swear.”

I scanned the security feeds. Bullets had torn away the gunwales, and were beginning to chew up the deck. Serene leaned out from behind the wall and fired several shots. A grenade landed on deck, and she dove for cover. It exploded, and bits of ceiling rained down into H3’s office.

“It’s getting hot up here,” Nash yelled. Two more barges approached the yacht as it drifted farther from shore. Nash fired at the first boat, but as he did, the second barge banked left. Several CFD soldiers rose from behind the gunwale and opened fire. Nash took multiple shots to the torso and one to the head. His body collapsed and fell over the side.

“Nash is down,” Serene yelled, and then unloaded a clip into the barge, hitting several of its occupants. The barge banked, out of control, swerving towards the yacht. “Fire in the hole,” Serene yelled, tossing a grenade over the side, into the boat.

The barge exploded, its shell careening into the side of the yacht, tearing a hole in the hull above the waterline.

That was the last of the CFD barges. As our boat floated away from shore, there was a relative silence and only the fire from shore continued.

As I turned back to the holovision, a chill ran through me.

H3 sat in the chair behind us. I turned around, ready to strike, but in the physical world, the chair was empty. H3’s avatar stood, and his voice resonated through the room’s speakers.

“Well hello gentlemen, and lady. It’s been a while.” H3

walked toward my image on the display. He put his augmented arm around my back. I couldn't feel it, but the hairs on my back rose as if I'd acquired a static charge.

"Kevin, what's happening?" Amelia said.

"I don't know," Kevin said, concentrated on the screen, waving through menus. "This doesn't make sense; he can't be ..."

"May I be the first to officially welcome you to Callisto," H3 said. "Unfortunately, your message on Cal-Net will not be broadcast today."

"Oh, shit," Kevin said.

"What's going on, Kevin?" I said, but he just shook his head.

A red notification flashed in the corner of the screen. I read, "Inappropriate content detected." Our message hadn't sent.

29

"We've got to get out of here!" I said. "If that message didn't transmit, those soldiers will keep shooting until there's nothing left of this ship!"

"Not so fast, John," H3 said.

Tayler ran into the room. "We're clear of shore, and they seem to be out of boats," he said. "But they'll have a 20-millimeter gun online in about sixty seconds that will tear this boat to shreds.

Avro followed him in, slinging his rifle back over his shoulder.

"Hello Avro," H3 said. "Who's your friend?"

Avro's skin went white, and he looked at Tayler.

"What the hell is going on?" Tayler said, looking at the holovision and seeing H3 superimposed onto the room.

"It's over for you, H3," I said. "We'll get a message out."

"You must have undergone some serious training, John," H3 said. "Last time we met, you were just an engineer. Unfortunately for you, there's one lesson it seems you haven't learned."

"And what the hell is that?" I said.

"Always have an ace in the hole."

Serene came down the stairs with gun drawn. Something had changed about her face and eyes. Perhaps it was from adrenaline, a heightened sense of awareness. She looked like a caged animal, planning to claw free past her keeper.

"Serene, don't move!" Tayler yelled, and reached for his side arm.

Serene pulled the trigger, firing three rounds into Commander Tayler. Two of them hit his chest, and the third struck him in the center of his neck. He fell to the floor, his gun clattering beside him. Serene walked over and kicked it away.

"Don't move," she said. Blood dripped from minor cuts on her face. Her stolen CDF tunic was slashed in several places, but her wounds looked minor and she didn't seem to be affected by pain. "Put your hands on the back of your heads."

She turned to Howard. "I'm sorry, but it's for the good of the colony," she said, and put a bullet in his head. He collapsed on the ground.

"Serene, why?" I said.

"She's working for H3!" Kevin said.

Serine pulled a second pistol from a holster. With one gun pointed at us, she pointed the second at the holovision.

"I need to talk to them, privately," Serene said to the holovision.

H3 put up a hand to protest. "Now, Serene, I need …" Serene fired one round into the projector and another into the camera.

"Oh my God," Amelia said. "You're a fucking Doomsdayer!"

Serene didn't respond.

"H3 will kill us if you turn us in!" I said. "No matter what happens, we're dead."

"I like you guys," she said. "And John, I like you a lot. Do exactly as I say and I can keep you alive."

"Why, just, why?" I said.

Serene motioned with her gun. "The Doomsdayers on Earth

made sure I would be on any mission to Callisto. The plan was to make sure when we arrived, the residents of Callisto would believe we were from the Communist Alliance."

"That's why you killed all those people," I said.

"It had to be this way," Serene said. "This settlement is everything to us. It is Utopia."

"That means you can never let us leave," Amelia said.

"The soldiers are going to come and take you away now. Throw anything with a NASA logo into the recycolizer." She pointed to the unit located near H3's desk. "Do it."

We ripped the Velcro NASA logos from our gear, throwing them into the machine that would instantly melt them down for filament.

"John, in the first aid kit, get the surgical anesthetic."

"You don't have to …" I began.

"Do it now, or when those solider arrive, I will *have* to kill you. The rest of you, lie down," she said. I looked in the bag and pulled out a handful of syringes.

Amelia scowled, giving Serene a look that said, *Fuck you.*

"It's okay," I said. "We can figure this out. We can't do anything if we're dead."

The four of us went down to our knees, and lay on the floor, hands still over our heads.

I pulled the safety tip off the first syringe, injecting Amelia just above the collar. She let out a silent scream. I lowered her to the floor as the drug reached her brain. Avro went next, followed by Kevin.

There was a pause. I turned and placed my hands on H3's desk. "Just fucking do it," I said.

"You'll be okay," Serene said. "H3 needs you. We need you. Whatever happens, I hope we can be friends again someday." I felt the prick, and the liquid shooting into my neck.

"Unlikely," I said, but as I slipped into unconsciousness I heard one last thing.

"Henry, this is Serene. Come get them."

From the top of the hill, Marie watched as a boat pulled up to H3's crippled yacht. A minute later, CDF soldiers hauled several bodies from the charred ship. When the boat moved away, there was an explosion and H3's boat sank beneath the surface. Marie had been on that boat this morning. *What would have happened if we hadn't evacuated?* she thought with a chill. The faces of her children flared briefly in her mind; the notion of them being orphans was unbearable.

She checked her watch, reading a Cal-Net notification, and letting out a sigh of relief. The immediate threat to the colony had been eliminated. She went inside the administration building. H3's door was locked so she knocked.

He came to the door and let her in.

"Have you heard?" Marie said.

"Oh yes," H3 said. "I've been following the counter attack on my headset."

"I've never seen you use VR," Marie said.

"I pulled this universal VR set from my spaceship and had a local tech network it into Cal-Net Social. It's kind of fun. You should try it someday."

"I've had enough virtual reality, thank you very much."

H3 turned on his room's holovision, now one of the only remaining units in the colony. They watched as reporters interviewed people by the shore. Other representatives from the council were giving statements. "This could be good for us, you know," H3 said.

"How's that?" Marie asked.

"When the Alliance realizes its reconnaissance team won't be coming back, it'll send another. But next time, it won't be so easy to fight them off."

"How is that in any way good?" Marie said.

"Unification, Marie," H3 said. "It will unite our colony like nothing has before. We'll go into full production of the spaceships, and we'll reactivate the universal constructor, doubling the size of the colony."

"And will this activity include a mission back to Earth?"

"Construction of the ship building facility has already begun. And, once we're confident we can match the firepower of the Alliance, well, I'm sure we'll be able to spare a ship for a reconnaissance mission to Earth."

There was a knock at the door. Marie turned and opened it, without waiting for H3's permission.

It was Hoshi.

"Hello, Marie," Hoshi said, and Marie nodded. Hoshi continued, "The army was a good idea after all. I wanted to thank you personally, but—" she said.

"Do you have the survivors in custody?" H3 said.

"We should have wiped them out!" Hoshi voice rose.

"They are prisoners of war," H3 said. "And we will not be conducting executions on my watch."

"Survivors?" Marie said. "I only saw bodies carried out."

Hoshi explained, "We killed several of the attackers, but some survived."

"What will you do with them?" Marie asked.

H3 said, "We'll be holding them in the *Mount Everest*. Locking them in VR. It will keep them sane until the war is over and we can send them back to wherever they came from."

"They came from Mars," Marie said. "Why do you sound like you don't know?"

“Mars, of course,” H3 said.

“I want to talk with them,” Marie said.

H3 looked at Marie, then back at Hoshi. “No one goes near the prisoners. Not even Hoshi or I. The Alliance has unmatched powers of persuasion, and we don’t want anyone to become … tainted.”

“I didn’t say I wanted to go near them,” Marie said. “I just want to ask them some questions. Besides you, these are the only people who didn’t come on the convoy. They might have information about Earth.”

Hoshi looked at H3. “She’s right, someone should conduct an interrogation.”

“I’ll do it,” H3 said. “I’ve met the CA face to face, and I believe I am the least corruptible.”

“Hand me your visor.” Hoshi said. H3 did as he was told. Hoshi put the visor up to her face, and typed a series of commands into the air, before handing the visor back to H3. Hoshi said, “I’ve given your visor permission to enter the Calli program. You’re the only one inside the Ring with access to virtual reality. Let’s keep it that way.”

30

I regained consciousness, but kept my eyes closed. Sunlight shone on the bed, and I pulled the blankets over my head. *What happened last night?* I asked myself, assuming I had been drinking, and this was the worst hangover in the world. Then I remembered. I opened my eyes, and threw off the bed sheets. The room was bright and cheery; the sun was shining through an open window, and I could hear birds chirping in the trees. I sat up, throwing my feet over the side of the bed. My head ached, but I managed to stand and walk toward the mirror. The man I saw in my refection wore shorts and a Hawaiian T-shirt.

"Shit," I said.

There was yelling outside the door. "John?" It sounded like Kevin.

I walked outside, finding myself on the second story balcony overlooking a pavilion with a stone fountain in the center. People walked around in normal clothes, going in and out of shops. Something was odd about the way they moved. It was as if they were all looking for something.

Kevin and Avro walked along the balcony toward me. Kevin was wearing his shirt with the F-35s. Avro wore a Hawaiian shirt, like me.

"Is everyone okay?" I said, walking up to Kevin.

"We're in VR," Kevin said.

"Yeah, I got that," I said.

"We're *locked* in VR," he said. "It's a prison."

Figures, I thought. I reached to my back, feeling around the port that held my body. I checked for the place where I figured the release would be. I found it, but the mechanism would not compress. I reached to my face, attempting to lift the glasses, but my hand couldn't get close enough. I must have been wearing a helmet.

Amelia burst through a door and stumbled into the light, using a hand to block the sun.

"What's going on?" she said. "Where the hell are we?"

"We're in VR," I said. "Based on the fact that they have our avatars, and that we're wearing what we wore in the Hawaii program, I'd say they have our ships."

"God dammit!" she said. Avro walked over and gave her a hug.

A spiral staircase descended from the far side and we walked down into the pavilion.

"It's the program Howard mentioned," Kevin said, "the one they used on the spaceships. These people are Turings."

"So, we're the only real people," Amelia said. "That's creepy."

"Howard never mentioned Turing computers," I said. A couple walked past holding hands. Two women approached each other from opposite directions; they hugged when they met, and held the embrace.

A teenaged girl sat beside the fountain, crying. She wore jeans, and a bright, long sleeved shirt.

"There's something very strange about this," Amelia said, and walked over to the girl. She sat down beside her on the ledge.

"What's your name?" Amelia said.

"Tanya," the girl answered.

"Tanya, I'm Amelia. Can you tell me what the matter is?" Amelia said. We sat down on the ledge and pretended not to eavesdrop.

"I don't understand," said the girl. "Don't you know?"

"No, we've been," Amelia paused, "away."

"There was pain," the girl said. "I couldn't breathe, no one could. I thought I was going to die. Our friends, from the other ships, they just … disappeared."

"What do you mean, disappeared?" Amelia said.

"I mean disappeared!" the girl said. "You can't just disappear like that. You can't just go back and leave us here!"

"Go back to where?" Amelia said.

"Back to reality, to the ship of course," the girl said. "You *are* from the *Klondike,* right?"

"Ah, yeah, we're from the *Klondike*," Kevin said, gabbing Amelia's arm. "Come on. We need to go."

"Don't go, please, don't go," Tanya said. "My family is dead and I have no friends."

"Stay here, okay?" Amelia said. "I'll be back."

The girl started to cry and held Amelia's arm.

Kevin turned to Tanya, looking her in the eye. "You are a Turing Computer – reset code four hundred and two."

"Oh, okay." Tanya smiled and immediately stopped crying.

"That was weird," Amelia said.

"It's a reset button," Kevin explained. He grabbed Amelia by the arm. "We need to talk."

We walked away from the pavilion and out into the main street that led down to the harbor. The VR world was almost identical to the Callisto colony, but with subtle differences. The dock in this version of Calli was made of concrete, in "real Callisto" it was made of wood, and in this version, there were more sailboats than speed boats.

"Howard said four ships left Earth, but only three made it," Kevin said. "What if only three ships left Earth?"

"And they turned off the Turings," I said. "Made it look like a disaster."

"The *Klondike* never existed," Amelia said. We took a moment to let the realization sink in.

"But they used the ruse to convince the population to create a baby boom, ensuring that any families on Callisto would be unable to return," I said.

"It's kind of brilliant if you think about it," Kevin added.

"It's sick!" Amelia said. "And Turing children? They created a being, Tanya, that thinks and feels, doesn't know she's a computer, and they gave her memories that say her family was killed in a nuclear war. Yeah, I'd say that's pretty messed up."

"We may be able to use this to our advantage," Avro said.

"How's that?" Kevin said. "We're locked in here. It's a closed system."

"These Turings *know* everyone who's on Callisto," Avro said.

"If we ask around, we may be able to get new information," I said.

"Do you think the Doomsdayers are listening to us right now?" Amelia said. "Spying on us?"

"Would we proceed any differently if they were?" I wondered.

Amelia shrugged. "Who should we talk to first?"

But before anyone could answer, Tanya came running.

"Amelia!" she yelled. "A man came to see me right after you left. He said to ask you to go to the theatre and meet him there."

"Where's the theater?" Amelia said.

"Back by the fountain," Tanya answered.

Amelia looked at us. "I guess we're going to the theater."

When we arrived back by the fountain, the girl pointed toward a row of doors that led into a fancy building across from the apartments.

The theater was empty. We walked between row upon row of red chairs. In the front of the room, eight stairs led up to the stage. The screen behind the stage was blank.

We stood near the stage, looking around. "This is strange," I said. "Why would a Turing lead us here?"

There was a clap, then another, and from behind a curtain walked H3. He was smiling, striding with smug confidence. Avro jumped up the stairs, throwing a punch at H3's head. His head lurched backward under the impact, but the smile remained.

Avro punched him in the gut twice then swung a right hook, and another punch to the gut. I ran onstage to join him. Amelia followed. H3's smile remained, unaffected by the blows.

Kevin stood at the top of the stairs. "He's not wearing a resistance suit!" he said. "You may want to stand back."

Kevin's warning came a moment too late. H3 turned to Avro, slugging him in the chest. Avro keeled over in pain. H3 turned to me, punching me in the face. He grabbed my hair, pulling my head down and into his knee. I stumbled backward, but H3 stepped forward, grabbed me by the belt, and threw me from the stage. I fell several feet, landing awkwardly on the chairs below.

H3 turned again to Avro who was just getting back up.

Amelia tried to hold H3 back, but he shook her off with ease, tossing her across the stage. She slid to a stop at the edge.

I peeled myself off the chairs, my back throbbing from the impact, but no more so than from any pain I'd experienced during our training.

H3 stopped in front of Avro, using a hand to brush imaginary dust off his shoulder. "If you've finished exercising, I thought maybe we could talk," H3 said.

"This makes no fucking sense," Amelia said.

"You've said that before," H3 said.

"Why the hell does everyone on this place listen to you?" she said.

"Why wouldn't they?" H3 said. "I'm a democratically elected official!"

"You helped kidnap thousands of people!" she said. "Why the hell would you have them believe the world was destroyed?"

"Think about it from my perspective," H3 said. "A few conspiracy theories here, a sprinkle of money there, and poof, I go away for a while, show up here and an entire society is waiting for me."

"I think you're taking too much credit," Kevin said.

"What about the other Doomsdayers, like Hoshi? How the hell did you get her to keep the conspiracy a secret?" Amelia said.

"Hoshi designed this utopia, she's proud of it, and she'll do anything in her power to keep it as the society she's envisioned. And, if you ask me, I kind of like it. And so does everyone else!"

I climbed back onstage and stood beside Avro. "Why haven't you killed us?" I said.

"Oh, there's nothing in the universe I'd love to do more!" H3 said. "You foiled my plans on Mars and I'll be damned if you do it again on Callisto. But as I said, I'm just a democratically elected representative. Having soldiers kill in battle is one thing, but an execution? I've yet to do anything quite so bold."

"You're still a murderer," Amelia said.

"Semantics, Amelia, Semantics. So, I'll tell you what I want." H3 paused, as if making sure we had his full attention. "I want a fleet of spaceships and a fully-integrated, defensive grid around this entire system."

"You have our JJs; what more do you want?" I said.

"Oh, I have the JJs alright. But I've castrated them. The CDF is under the illusion that the spacecraft were stolen from me on Mars. I have my pilots training in them right now. But since I don't need anyone escaping, we removed the JJ's nuclear reactors. We'll use them to power the defensive grid we're about to build, that is, when I get permission from the council. The council will love my plan for nuclear-powered rail guns surrounding Callisto, preventing any ships from getting in. And, thanks to your gruesome and merciless attack on our fine colony, I'll have all the support I need."

"A fleet of spaceships and a defensive grid?" Amelia said. "You really want to cover your ass."

"Here's the thing. These people are stuck here. Even if they find out Earth is still around, they're stuck here until they rebuilt the spaceships, fuel them, get them back in orbit, etcetera, etcetera, etcetera. And, on top of that, a third of this colony is under the age of two; they'd never survive Earth's gravity. It's not going to happen. We've jammed all radio communication in and out of the Jovian system. Hoshi thinks that's enough. But I disagree. I won't let another convoy like yours land, and I won't have anyone finding out the truth about me."

"Everything is always about you," Amelia said.

"What do you want from us, specifically?" I said.

"You know how many aerospace engineers there are in this colony? None! We kidnapped thousands of able-bodied adults, a lot of great mechanics, but not one of them knows how to properly design a freaking spaceship."

"Then who built the convoy?" Kevin said.

"You mean those off-the-shelf mining vessels? They stuck an inflatable module on top for a habitat. That was LEGO, Kevin. I need fighter craft, interceptors, everything those JJs are and more."

Kevin looked at us and shrugged. H3's story was believable enough.

"John, you were on the design team for the orbital ring, Destiny Colony, were you not? And Kevin, I know you've designed several aircraft. Avro is probably the best pilot in the solar system; who better to test a new spacecraft? And Amelia, you're well versed in weapons systems, I'm sure you can lend a hand, too."

"Not sure we can help you …" I said. "Building weapons for our worst enemy."

The screen behind H3 came to life.

"Here are the criteria. I want a single pilot, short range spacecraft, capable of achieving orbit of Callisto, but incapable of interplanetary flight. I don't need my soldiers going AWOL. The spacecraft will be powered by hydrazine and fuel cells. No VR interfaces, good old stick and rudder."

"You really think we're going to help you?" I asked.

"Oh, I think you will. You see, if you help me design my spaceships, I won't kill your wife and son."

I locked eyes with H3, my breathing on hold. *Does he mean …? He can't mean …? What the hell!* H3 held my burning gaze, and let his statement sink in for a moment, but it didn't make any sense. Marie and Branson were dead!

"*My* wife? *My* son? What are you talking about?" I demanded.

"Your wife, Marie, is a lovely woman," H3 said. "She's here on Callisto."

I reeled back in shock, my heart thumping painfully.

Behind H3, the specifications disappeared, replaced by a video of Marie on a run. She wore a bright blue windbreaker and track pants. Her curly hair was tied back in a bun. She used to always run early in the morning, before it got too hot.

"It's a fake," I said. I was struggling to breathe, hating H3 more than ever. Deep down, I knew this wasn't a deception, it was blackmail.

"Let me tell you a story about your wife. She's quite a brave woman. After the impact, a Doomsdayer named Hoshi rescued Marie and Branson on Highway One in Marin County, California. We targeted her, John, long before the *Bradbury* disaster. Marie was the world's expert in population sustainability. Marie became the colony's lead expert in genetic diversity. She used the *Klondike* disaster to inspire all of society into create a new generation."

"This is lunacy," I said. My legs shook, and my knees gave way. I sank onto them.

"Oh, there's more. You see, Marie found me shortly after I landed on Callisto. I brought her into my home. See, here's a picture of us together." On the screen was a picture of Marie smiling with H3 on the steps of what looked like a government building."

"This is fucking insane," Amelia said. "Why are you doing this to him?"

H3 continued, "I was shaken up after my experience on Mars, but Marie not only convinced me to rejoin society, but became my personal assistant, helping me get elected as a representative for this fine colony."

"I don't believe it," I gasped. "Marie would never help you."

"And why not?" H3 said. "I'm a hero who tried to save Mars from the Alliance, and now I'm trying to save Callisto. How about this?" H3 went to the next slide; it was a video. Branson ran through the grass. He was so big now. He ran towards H3, H3 scooped him up, and then smiled at the camera.

It was too much. I bent over my knees as if punched in the gut, and cried out like an animal in a trap. "I don't believe it, I don't believe it," I said, my whole body shaking. Amelia walked over and put her hands on my shoulders.

"Oh, but there's so much more to tell you, John," H3 said. "You see, everyone on Callisto has a duty to help grow the population, grow it to a point where it's sustainable." A knot of dread tightened in my stomach and I felt the rations I'd eaten churning. "You're going to love the next slide. It's family photo, of sorts."

I craned my neck upward, despite the pain. The photo on the screen was of Marie, holding a little girl of perhaps two.

"I'd like you to meet Lise."

"No fucking way!" I yelled, and sprang from the floor to run towards H3. I hit him, punching him until my hands burned inside the resistor suit. His avatar swayed with each blow, but nothing I did would cause him any pain.

Avro caught my shoulders and held me hard, pulling me away from the avatar. My friend's voice barely penetrated the red haze that filled my brain. "It's not his child, John. He hasn't been here long enough."

"Design my spaceships," H3 said. "You have one week." I looked at the others, and realized they were watching me, assessing my reaction.

Before we could say another word, H3 reached up to his face, removing the headset. His soulless avatar turned, went behind the curtain, and disappeared from existence.

31

We walked out of the theater into the pavilion. My legs felt far away, and my brain was numb and blank.

"What do we do?" Amelia said.

"What can we do?" Avro answered. "We build his ship. He's trying to cover his ass. If we don't build it, someone else eventually will."

"Kevin, can you sabotage the design?" Amelia said.

Kevin shrugged. "There's not much to sabotage that the workers won't catch when they assemble it."

"I'm going for a walk," I muttered. "I need to be alone for a while."

I left the pavilion, and headed toward the water. The breeze blew through the sails of the boats docked on shore, and ankle-high waves lapped at the rocks. I tried to clear my mind, concentrating on the sound of my footsteps on the boardwalk. *Is Marie really alive?* I asked myself. *Was that really Branson?* A powerful tide of hope flooded me, overpowering the hatred I felt for H3.

The other people on the boardwalk looked as if they were doing the same thing as me, dwelling on things they didn't want to accept. About a kilometer and a half from town, I stopped at a bench and sat down.

A woman in jogging clothes sat beside me. She opened a bottle of water from a pouch on her belt and took a sip. She sat staring over the waves.

"I can't believe they're gone," she said. "I think they're out there, somewhere. Most people don't, but I do."

I didn't answer, I just stared.

"You must have lost someone," the woman said. "I can tell you miss them. I lost someone, too."

I nodded, as if agreeing. Then I looked at the woman, who had tattoos on her arms. Words written in a script I didn't recognize.

"Well, it was nice to meet you, I'll be on my way now," she said, getting up and clipping her water bottle to her hip. "My name's Lise. I'll see you around."

"Wait," I said. "What's your name again?"

She looked at me with skepticism, as if trying to decide if she knew me. "Lise," she repeated.

"Lise. Is that a common name?" I asked.

"No, not really," Lise said, sitting back down, and holding eye contact firmly. "Why?"

"There is this kid, a baby girl, her name is Lise," I said.

"Is she on this spaceship?" Lise said.

What the hell am I doing? I thought. *I'm sharing an experience with a computer program that thinks it's real.* "Ah, no, she's not," I said.

"She's missing then, I'm so sorry," Lise said.

"No, I don't think she's missing. In fact, I'm not convinced she exists at all," I said.

"That's unusual, not knowing if someone exists," Lise said.

"A lot of things today have been unusual," I said. "For example, the people here, think …"

Lise stood up, and turned to face me, as if all of a sudden I wasn't who I'd said I was. "You're not from the *Klondike,* are you?"

"No," I said, not thinking that there might be consequences for revealing who we were.

"I arrived on a separate ship," I said.

"And you found us floating through space cut off from the others! Tell me what happened to the other ships? Are the people okay?" she asked.

"Yes," I said. "The people in the other three ships are fine."

"Oh, I have so many questions!" Lise said with a smile.

"Later," I said. "I need to go." I got up and began to walk away.

Lise's smile was replaced by a look of confusion. "What's your name?" she asked.

"John," I said over my shoulder. "John Orville."

"Wait," Lise yelled. "By any chance is your wife's name Marie?"

I stopped. My gut wrenched and tears formed in my eyes. I turned and ran back toward Lise, grabbing her by the shoulders. "What do you know about Marie?"

"She was my best friend," Lise said. "We worked together at the Center for Genetic Diversity."

"You worked together?" I said. "Tell me everything!" Was this some elaborate scam or had H3 told the truth? I didn't care; I'd let myself bask in the illusions.

"She was on the *Mount Everest*. A ship from Tibet," Lise said. "Hoshi, a Doomsdayer, rescued her and Branson from California."

"You know Branson?" I said.

"Only through Marie, I never went to the nursery," Lise said.

"The nursery," I repeated, not sure I was asking a question.

"The younger kids aren't with us in VR," Lise said. "We visit them in the nursery."

"Did Marie have a daughter?" I asked.

"No, just Branson," Lise said. "But she was willing to have another child. I promised to help her pick out a donor, but she wanted to wait."

"What's today's date?" I asked.

Lise checked her watch. "It's a Sunday, January fourteenth."

"What's the year?"

"That's a silly question," Lise said. "How do you not know the year?"

"Just answer me, please."

"It's 2072," Lise said.

"2072," I whispered. "Shit, that's three years ago."

"What are you talking about?"

"Never mind, yeah, 2072, right. Do you know about a man named H3, or Henry Allen?" I said.

"No, doesn't ring a bell," Lise said.

"A business man, he lived in New York. President of Red Planet Mining?"

"Sounds familiar," Lise said. "Was he a survivor?"

"He was, ah, on Mars. The other ships are the *Mount Everest*, *Victoria* and *Melbourne*, right?"

"Right," Lise said.

"Listen to me, no matter how strange this sounds," I said. "Those ships have already arrived on Callisto."

"Impossible. There are months left before we arrive."

"They arrived almost three years ago," I said, sensing fear in Lise's eyes, and I wondered how her program would handle the paradox. "This program, the one you're living in, it has been offline for years."

"I don't understand," Lise said.

"That's because you are part of the program; you are a Turing computer." Then I remembered what Kevin had said about the reset button. I said, "Reset code four hundred and two."

"Oh well, that makes sense then," Lise said.

"So, what now?" I said. "I just told you that you aren't real."

"I'll just add that fact to my memory. Thank you, John."

"Lise, we've been imprisoned here, in this program, by a man named H3. He's threatened to kill Marie if we don't design a spaceship for his military—"

"*Kill* Marie?" Lise said. "Then design the damn ship already!"

"We will, at least, that's the plan, but I need to get out of here, I need to save Marie. That means getting out of VR. Do you have any idea how to do that?"

"As you say, I'm a Turing, I don't know what I can do to help. But there is a manufacturing facility near the Ring's wall. If you are designing a spaceship, that's where you'll want to do it."

"Thank you, Lise," I said. "If you come up with any ideas, that's where I'll be."

32

The others had already found the facility that Lise mentioned. I arrived to find my three friends designing the spacecraft on floor-to-ceiling holodisplays. Kevin had printed a frame so we'd have something to touch, but it looked like a school bus without the body, seats or engine. They stopped working when I showed up, and Amelia walked over and gave me a hug.

"I found a Turing that knew Marie," I said. "Unless the place is designed to mess with our heads, it validates H3's story."

"I'm sorry," Amelia said.

"Her name was Lise, like Marie's baby," I said. "My guess it that Marie named the child after her."

"Did you tell her about Marie?" Amelia said.

"I did, and I told her she was a Turing. I also asked if she could help us in any way. She said she didn't think she could."

"I don't think the Turings are going to be much help," Avro said. "Not until we start developing the manufacturing process anyway. I'm just happy they're staying out of the way."

Avro nodded to a panel with a 3D image of the blank frame. "She's all yours, boss." I started by programing in the orbital and gravitational characteristics. This would determine the size of the fuel tanks required. We'd even need to design the rocket engines from scratch.

Kevin was designing the spacecraft's body. He stood in front of his display, looking frustrated. "What's going on, Kevin?" I said.

Kevin said, "If I had my files, or access to Space-Net …"

"Well, it's a spaceship, so it doesn't need to be shaped like anything. It can look like the lunar lander for all we care," I said.

Avro stepped out from behind a display. "No, that won't do. This thing is a fighter," he said. "It needs a cockpit with good visibility so the pilot can point the guns at the enemy." Avro pointed with two fingers on each hand, as if aiming a set of aerial cannons.

"If you want visibility, place the cockpit off the bow, like a Romulan Warbird," I said.

"Romulan?" Avro said.

"You know, from *Star Trek*?" I said. Avro just shrugged. "Never mind."

"I think it needs to be like an A Wing," Kevin said. "You have seen *Star Wars*, right?"

"I think you're having fun with this," Amelia said, leaning out from behind her display.

"Okay bomb expert, are you making any progress?" Kevin said.

"I've been thinking about the targets, ships like our JJs, drones or massive transports. In a vacuum, all you need to do is punch a few holes and that's a hit."

"Then just use whatever the JJ's used," I said.

"The JJ's primary weapon was a rail gun. The projectiles were flat disks ejected from a coil. But the rail guns are nuclear-powered. The fuel cells in your design won't produce enough power."

"What's your solution?" I said.

"Bullets. We fill the chamber from the O2 tanks. I just need to figure out the caliber of the bullets used by the CDF."

"They looked like twenty cal," Avro said.

"Yeah, that's what I've assumed," Amelia said. "I've gone with that for now. I'll leave the design open if we need to make a change."

"How are we going to test this thing?" Avro said. "Should we have Kevin print us a flight simulator?"

"Flight simulator?" Kevin said. "We're *in* a simulator. We'll just move it to the airlock."

"That means we'll need a vehicle to tow it with," I said.

"On it," Avro said, and began working at an adjacent console. "There's no shortage of tractors in the database."

There was a knock on one of the panel displays.

We froze, and H3 walked in from around a corner. He paced about the factory floor, looking at our meager frame and at each display in turn.

"What do you want now?" Avro said.

"Good afternoon, gentlemen and lady," H3 said. "I wanted to confirm that you had accepted my offer. It looks like everything is in order here."

Avro gestured to the ship's frame and to the design we'd mapped out on the displays. Kevin had sketched out an A Wing from *Star Wars*.

"Very nice," H3 said, looking closely at Kevin's design. "Very original." He walked over to Amelia's console. "Interesting," H3 said and then passed his wrist over the consul and swiped the weapon's design onto his watch.

"Impatient much?" Amelia said.

"A gift for our military commander," H3 said.

We heard a knock, coming from H3's avatar. "Duty calls," he said, lifting his hand to his head. The avatar backed away from the console and walked out of the room.

"Wait!" Amelia said, but once he was out of view, the avatar was gone.

Marie walked into H3's office with Commander Yamamoto. H3 set his headset in a drawer, closing it before standing to greet his guests.

"Marie, Commander," H3 said. Marie took a seat across the desk from H3 and took out her notepad. Yamamoto bowed and took the other seat.

"I have something for you," H3 said. "Something I've been working on." H3 reached to his left wrist, and used a gesture to transfer a file to the commander.

"A weapons system?" the commander said.

"For use in a vacuum," H3 said. "Just a little something I've been working on."

"Impressive," said the commander, pulling a screen from his watch and inspecting the document. "I had no idea you had such a talent for computer aided design."

"Please, install this on the JJs you apprehended; to my knowledge the rail guns won't work without the nuclear reactors."

"Very well, Mr. Representative," Yamamoto said. "But there is something I need to request of you. It is my responsibility to ensure the defense of this colony. That includes gathering intelligence in any way I can."

"You want to talk to the prisoners," H3 said. "I assume you asked Hoshi and she denied your request. Now, you're coming to me."

The commander nodded. "Hoshi has been most unhelpful."

"Those men have been to Mars. There is nothing they know that I do not. If you wish to interrogate someone, you may interrogate me."

"Representative," the officer said, "there are things that only a trained military officer will understand. I'm going to interrogate the prisoners, Mr. Representative. Letting you and Hoshi know is, simply, professional courtesy." Yamamoto got up, bowed again, and turned to leave.

"Excuse me, Marie, I have to talk some sense into the commander." H3 followed him out.

Marie was left alone in H3's office, and knew exactly what she needed to do. *If I'm going to find answers, now's my chance.* She looked at his desk and reached into his drawer, pulling out the VR set. She placed the visor over her eyes and found herself in an augmented view of H3's office.

The glasses scanned her eyes, matching her profile from Cal-Net Social. The machine cross-referenced her profile, found her avatar in its database, and loaded it.

Marie said, "Menu, select program, Calli."

The view of H3's office disappeared as the system transitioned from AR to VR. Marie's avatar materialized alone in the theatre.

In reality, she stood, and pushed back H3's chair to give her some room to move. The units sensed her hands and arms but, unlike during her time in Calli, there was no resistance.

Marie leaned forward and her avatar began to walk. She straightened and it stopped. The movements Marie needed to use were subtle; directing her avatar was like riding a horse and directing it with body posture, weight shifts, and leg pressure, instead of by the reins.

There were mirrors on the wall and Marie saw her reflection. Her avatar's hair was shorter and had less grey. She felt as if she'd stepped into the past, and in a way, she had.

She strolled out into the pavilion. It was either evening or morning in Calli, she couldn't tell which, but several people wandered through the streets. *Are these the prisoners?* She walked up to a white man in a suit. He didn't *look* like a prisoner. "Hello," Marie said.

"Can I help you?" the man said in an American accent.

"I recognize you," Marie said. "You were on the *Klondike*."

"You mean I *am* on the *Klondike*," the man said. "And you're not?"

"Oh my God," Marie said. "You're a Turing computer."

The man turned and walked away.

Marie remembered. "Lise!" She ran to Lise's apartment and banged on the door, but Lise wasn't there.

Back at the pavilion, Marie looked around again, muttering, "Where could she be?"

Marie walked up to a middle-aged woman, asking her for the date and time.

"Monday," the women replied. "Five-fifteen."

"Thanks," Marie said. *Lise must be on a run,* she thought. Marie jogged down to the park and looked around. Several people were walking about. She saw her friend loping along one of the paths.

"Lise!" Marie said, then remembered to keep her voice down. Her real body was in a government building, after all.

Lise stopped. "Marie? Is that you? You were gone, oh my God! I thought I'd never see you again."

Lise hugged Marie's avatar. "But you're a Turing," Marie said, sounding shocked.

"So I've been told," Lise said.

"Do you know anything about the prisoners they're holding here?" Marie said. Lise nodded, opening her mouth to talk, but Marie interrupted. "Can you tell me about them? Have you talked to them?"

"Yes, in fact, one of them is your husband," Lise said. Marie froze, her brain barely processing what she'd heard.

"My *husband*?" Marie said. *Is this some sort of Turing*

glitch?

“Your husband, John.”

"Lise, that doesn't make any sense."

"Come, I'll show you. They’re in the factory designing a spacecraft of some sort.”

“A spaceship,” Marie repeated. Fog seemed to fill her brain, and her mouth felt wooly when she spoke. *How could John possibly be here?*

Lise explained, “H3 was looking for someone to design a spaceship for him. He was frustrated that no one in the colony knew how. We must hurry. John has something important to tell you, something you should discuss together.”

Marie stumbled after Lise, her heart alternately speeding up with hope and slowing down with the dread that Lise was mistaken.

They reached the factory and went inside, making their way to the assembly room. Two men, one Indian and another of medium complexion, stood together, unloading panels from a printer. The panels were curved, and looked like fenders on a car. They hung from a crane, which the men brought over to the spaceship Lise had mentioned.

One frantic glance showed Marie that neither man was John. “I don’t see him,” she choked.

“He must be here,” Lise said.

“They’re ignoring us,” Marie said. “They believe everyone else in Calli is a Turing.”

“Hey,” said a woman, who appeared to be working on some sort of gun. “Can we help you? Hey, I’m talking to you!” The woman got up and walked toward Marie.

But Marie’s eyes were riveted elsewhere. Movement on the port side of the ship. A man’s body. Marie’s eyes widened and her mouth opened in a silent shriek of recognition, shock, and joy. The man climbed out from under the ship, walking around it, inspecting it.

In H3's office, Marie's body lurched with urgency. Her avatar walked forward painfully slowly, or so it seemed to Marie. She wanted to hurl herself into her husband's arms; to be there! "Faster, dammit!" Marie urged. As her avatar got closer. Marie whispered as loudly as she dared, "John!"

The port fuselage seemed to be fine; it had been slightly dented on the tow into the airlock, but we had printed a new component. My eyes drifted over the ship. *So far, so good. Good enough to satisfy H3. And keep Marie and Branson alive.* My heart gave that familiar squeeze of bittersweet joy when I thought of them. I was still trying to get used to the fact that they were here.

There was movement from the corner of my eye, and I straightened up. Amelia was shouting at—a Turning? An avatar of—*Marie*? I opened my mouth to call her name but no sound came out. I leaped forward, feet slapping the floor hard, arms pumping. Finally, sound burst from me. "Oh my God, Marie!" I sprinted across the facility.

We were a dozen feet apart when Marie stopped, her expression blank. I heard yelling. Suddenly she turned, heading back in the direction she'd come.

"John," Kevin yelled. "She's jacked out."

"No!" I cried. "Marie!"

Marie's avatar dropped out of sight behind a wall. When I turned the corner, she was gone.

The visor was ripped from her head, and Marie came face to face with H3.

“What did you see?” H3 yelled. “What the hell did you see?”

“I saw my husband!” Marie shrieked, her voice strained with emotion. “Why is my husband in Calli? Tell me he isn’t real! Tell me he’s not one of the prisoners! Tell me you’re not using me!”

H3 leaned close, whispering in Marie’s ear, “Be very careful what you say from now on, Marie.”

“You bastard!” she yelled, and began to cry, slamming his desk with her fist.

H3 continued to whisper, "You say a word of this to anyone and you'll die. Tell anyone about what you just saw, and they will die.”

Marie pushed H3 out of the way. She rushed out of the door, blinded by tears, her head spinning, and into the artificial sunlight. H3’s personal CDF bodyguards watched her impassively. She half expected them to stop her, but they didn’t. It was H3 who followed her, jogging to catch up.

“Get away from me!” she screamed.

He ignored this. “It’s not me who’ll do it, Marie,” he said as they descended the steps of the government house. “Hoshi has spies. I don’t know how many, two, maybe three, but they’re here, and they work for her.”

"My God,” Marie said, turning to face H3. “You son of a bitch! The world still exists, doesn't it? We were fucking kidnapped! This is part of some massive deception!"

H3’s composure didn’t waver. “There’s still a chance you can get out of this, live your life as if it never happened. I promise I won’t tell.”

“You lied about everything. There is no Communist Alliance, not in space. My husband is in there, which means that …” Marie paused, “means he became an astronaut, that he came to rescue us! Why did you order the soldiers to attack them?"

"Marie, this deception keeps us alive."

"This deception keeps you in power!" Marie said. "I demand to speak to my husband."

"That's not going to happen," H3 said.

"The hell it's not." Marie started to run, leaving H3 alone on the street.

⚡

She jogged toward the place where they'd first entered the colony, near the mudroom where Malcom and Huey had tried to contact the lunar base.

Marie found Malcom working in the city's utility building, "You," she said fiercely, skewering him with her glare.

"Whatever it is, not now. We've just been ordered to start working on this defensive grid and an assembly line for some new type of spaceship, and I've got no time—"

"It's about that new ship," Marie said, hoping to grasp his attention.

"Do you know where they're getting the designs for it?"

"Yes," Marie said.

Malcom grabbed a recently printed coil of wire and walked out the door of the utility building, and into a newly constructed manufacturing facility on the Ring wall. Marie chased after him. Inside the building, workers, some in CDF uniforms, some in coveralls like Malcom's, were busy setting up assembly lines.

"I'm going to tell you something. Once I do, everything will change," Marie said as she jogged to keep pace with Malcom's stride.

"Okay, what?" Malcom said, placing the coil on a table.

Marie grabbed his arm, dragging him toward a long hallway that bordered the Ring. One side of the hall was lined with EVA equipment, spacesuits, tow bars and battery-operated tools. The other wall was lined with airlocks spaced equally apart.

Marie made sure they were alone. "I've talked to my husband."

"What are you talking about? When did you get married?"

"I didn't *just* get married, my husband is from Earth, and he's alive! Earth is fine! And I have proof. The prisoners aren't from the Alliance. They're from NASA."

"Shit! Marie, that is officially the most bat-shit crazy thing I've ever heard. What possible evidence could you have?"

"You're going to help break them out," Marie said, still holding Malcom by the arm.

"I don't believe you," Malcom said.

Behind them, the sound of a gun cocking. "Believe her."

"Who are you?" Malcom said, turning to address a blond woman in a CDF jumpsuit. She tossed Marie a set of zip-tie handcuffs.

"Put these on him. Get moving. Scream and I'll shoot." The mysterious woman led them down the hall. She pointed the silenced end of her pistol at Marie. "Tighten them."

"I'm sorry," Marie said, reaching toward Malcom's back and pulling on the ties.

Marie glanced through the airlock portholes on their right. Inside were rovers designed for Callisto's surface.

"Why are you doing this?" Marie said.

The woman drew Marie close to secure the zip ties around her wrists.

"H3 told me who you are and what you know," the woman said. "You used to work at the CGD, right? It *was* your job to ensure the sustainability of the colony. Now that job falls on me."

"But …" Marie said.

"You really want to know who I am, bitch?" the woman said. Marie didn't answer. "My name is Serene, and I'm your husband's girlfriend."

Marie turned to look at the crazed lady, noticing her blond hair tied tightly in a bun. *How could he possibly fall for a woman like this?*

Marie's body tensed against this fresh pain; one more betrayal.

Serene adjusted her gun, pressing it against Marie's face, forcing it forward.

Serene ushered them down further down the hall. They stopped at an empty airlock. The woman punched a button on the wall and a red light flashed above the door.

Marie looked around for some way to escape, and wondered whether to scream. At the end of the hall, behind a double door, printers wailed, and men in welder's masks worked with heavy machinery. No one would hear her no matter how loud she was.

"What are you doing with that airlock?" Malcom said, as the door hissed open. But the woman shoved the gun into his ribs and forced him forward.

"Do what I say or I'll put a bullet in Marie's head." Serene held Marie by her shirt at the back of her neck, pointed the gun at her head, and then pointed the gun at Malcom.

"There's no way I'm going in …" Malcom cried.

Serene pulled the trigger, firing a silent round into a panel beside Malcom's head.

"Get inside," she said, pointing the gun back at Marie.

Malcom stood obstinate in front of the open door, hands tied securely behind his back.

"Fine," Serene said, "have it your way." She held Marie back and kicked Malcom in the chest. He flew backward, landing on his shoulder, groaning in pain, and struggling to get up.

Marie screamed as Serene hit the button and the hatch began to close.

"You're about to have an unfortunate accident," Serene said.

Out of the corner of her eye, Marie saw a flash of red, and heard a crunch.

Spinning around, she saw Huey. Then she looked at the wrench embedded in the back of Serene's head.

"What an unfortunate accident," Huey said.

Serene slumped to the floor, smashing her dead face into the airlock's hatch.

"What the hell, Marie!" Malcom said, and Huey rushed into the airlock to help his friend. "That woman was going to blow me out the airlock!"

"I'm telling you, we're living a lie," Marie said, as Huey opened a pocketknife and cut the ties that held their hands.

"We were kidnapped from Earth to form this, whatever this is; this drone-free, AI- free society."

Huey looked at Malcom, then back at Marie, and pulled his wrench from Serene's head. He opened an access panel on the wall, revealing a cylindrical vat filled with filament. He dropped his wrench into the vat.

"Help me with this, would ya?" Malcom said to Huey, grabbing Serene from under the arms. "This never happened."

The two men hoisted the woman up, tipped her head backward into the vat, and dropped her in. Her body sank into the goo and Huey closed the panel, hitting a button on the wall marked "Recycle."

Marie slumped to the ground, bile rising in her throat. It was all too much to take in. "This is why H3 stopped your transmission."

Huey lowered himself, and put a hand on Marie's shoulder.

He'd returned to his silent, contemplative self.

"How could they hide this from me!" Malcom said. "I was the prime radio operator on our ship." He paused, and then said, "Oh my God. The communication team, all the radio experts, my boss, they were all on the *Klondike*, and now they're dead! The Doomsdayers covered it up by murdering all those people!"

"Dammit, Malcom, the *Klondike* didn't exist! Those people we watched die, they were Turings. I talked to several of them *today* using H3's visor."

Malcom went silent, suddenly feeling embarrassed. "I'm an idiot," he said.

"You're not an idiot," Marie said. "You're one of the smartest people in the colony. If we can free the prisoners, free my husband, we can tell the colonists the truth. The truth is the only thing that will keep us safe now. That's why I need you to believe me."

"I believe you," Malcom said. He reached over and hit the access panel on the adjacent airlock. Additional lights flickered on, illuminating the airlock's oversized interior. Inside was a rover designed for moving workers outside the Ring habitat. The rover's hotdog cabin rested on six large titanium tires. Oval windows lined the sides while two tinted windshields came to a point at the front.

"What are you doing?" Marie asked.

"I have an idea," Malcom replied. "Huey, keep watch from the observation deck, we may need your help." Huey nodded, and turned to leave.

"Marie, come with me," Malcom said. "We're going to get your husband."

The rover's rear hatch doubled as a ramp, and Marie and Malcom climbed up. Malcom hit the switch to close the door. They crossed a platform lined with spacesuits, and six rows of bench seats. Malcom sat in the driver's seat and began throwing switches. Marie sat on a bench seat behind him.

Malcom looked at a camera on the airlock wall and gave a thumbs-up.

The airlock depressurized, and opened to the vacuum of Callisto's surface. It was nighttime on this side of Callisto, and would be for the next eight days, but Jupiter glowed bright in the sky. The giant planet illuminated their surroundings, casting hard shadows.

They backed out of the airlock and turned around. The road ahead was banked with slag, like black snow. They continued past the slag and onto a tarmac where the mothballed *Melbourne, Victoria,* and *Mount Everest* loomed.

The *Mount Everest* looked different than the others; its service module was supported by scaffolding, and the Bigelow module had been inflated.

"The CDF inflated the ship for use as a brig. Prisoners are locked in permanent VR, catheters and everything. That way, they don't need guards."

He pulled up to *Mount Everest*'s airlock, extended the docking port, and pressurized it. Malcom got out of his seat. The port extended from the side of the rover. He opened the interior hatch, and inspected the door on the ship. He tried to spin the hatch, but it was locked.

A red light flickered on a panel located on the hatch.

"Problem?" Marie said.

"Working with the CDF is a pain in the ass. Mainly because they're green, and incompetent, at least from a technological standpoint. Huey and I have a way to work around pretty much everything."

"I see," Marie replied.

Malcom tapped his wrist.

"Huey, the *Mount Everest* is locked. See if you can override the CDF security protocol."

"On it," Huey said. There was a pause, and an alarm

sounded. “Wait, something’s going on here, they’re evacuating all non-CDF personnel out of the building.”

“Huey, just get the airlock open, whatever it takes!”

“There’s someone here, I’ve got to …” Huey never finished his sentence, but Malcom’s watch produced the sound of two cracks, like boards slapped together. Gunfire.

“Huey!” Malcom yelled into his phone. “Oh, God!”

“They killed him,” Marie said, almost matter-of-factly. She was so far into shock that now everything seemed routine. “They killed him!” Her hands began to shake.

Malcom was breathing heavily and rested an arm on either side of the docking port. “They’ll pay,” he said and closed the interior hatch. “They’ll all pay. Have you ever worn a spacesuit?”

“No,” Marie said.

Malcom retracted the docked port. “Well, neither have I. But I took an emergency EVA course, once.” He threw the spacecraft into reverse, backed it up, and turned until the front of the rover faced the *Mount Everest*’s inflated module.

“We’re going in,” Malcom said.

“Through the side?” Marie said, looking up at the giant marshmallow of a spaceship.

“Yup. They’ll lose oxygen. But their suits will stay pressurized until we get them onboard. Are you with me on this?”

Marie nodded.

Malcom got up from the driver’s seat, and began pulling on a suit. He snapped the helmet over his head and activated the comm unit on his wrist. Marie pulled a suit over her clothes, and Malcom checked the seal on her helmet.

He returned to the driver’s seat, pushing it back to accommodate the backpack. “Ready?” he said.

Marie sat down, reached for her wrist, and switched on her spacesuit’s comm. “Do it.”

Malcom gunned the accelerator and they launched forward. The rover struck the *Mount Everest* at thirty-five knots, tearing the canvas hull like a rag. The blast filled the surrounding vacuum with dust, which fell to the ground in exactly two point five seconds. The *Mount Everest*'s hull began to collapse as the tear widened, ripping to the zenith. Anything that wasn't fastened down blew onto Callisto's surface.

It took less than five seconds for the module to depressurize. But for those few moments, the sound of air rushing past the rovers was deafening. When it stopped, the only remaining noise was the servos inside the rover.

Headlamps above the windshield illuminated the interior of the spaceship. The prisoners hung from the scaffolding, about twenty meters from the rover's entry point. Without oxygen, they struggled for air like men on the gallows, hoods and helmets fastened tightly around their faces.

Malcom pulled the rover as close to the prisoners as he could, and dropped the rear hatch. He grabbed an emergency ax as they rushed down the ramp.

"They're suffocating!" Marie said, as they reached the prisoners.

Malcom lifted the ax, smashing the locking mechanism the held the first prisoner to the ship.

The body dropped from the scaffolding onto the floor. Water sprayed from the tubes, sublimating instantly into gas.

"These are military grade VR suits. They'll stay air tight until we pressurize the cabin." He moved onto the next prisoner, cutting loose their cables.

Marie dragged the first prisoner underneath the arms, pulling the body up the ramp. The prisoner's feet kicked aimlessly, his arms grasping for his neck which he, or she, couldn't quite reach through the helmet.

The second body lay limp on the ground like a doll. Malcom

picked the body up at the hips, heaving it over to the ramp. He lifted the body, tossing it into the rover. The figure arched in the mild gravity, sliding to a stop on the floor.

Malcom cut down the third prisoner. The body flopped lifelessly onto the floor. Marie dragged it up the ramp by the feet. This one was lighter than the others, probably the woman she had seen while in VR.

Malcom cut down the fourth and final body and tossed the ax aside. "Clear the ramp," he yelled, and pulled the last body up into the rover. He hit the hatch controls and the door clicked shut.

Air shot from vents along the ceiling. A green light flashed and Malcom tore off his helmet. Marie did the same.

He ripped off the first suit, revealing a man of Indian descent, who slurred what sounded like curse words in a Hindi dialect.

Not John.

"You sure these ain't people from the Alliance?" Malcom said.

"Shut up, and get their suits off," Marie said, pulling the hood and helmet off the next person. Marie started to panic, knowing that if they didn't get the suits off in time, John would be dead.

The second person, a woman, coughed, and sat up, eyes wide, then turned over on her side, gasping for air.

Not John.

Malcom reached for the third prisoner, tearing off the hood to reveal a man with a chiseled jaw and black hair.

John? Not John. Panic kicked Marie in the solar plexus, winding her.

The third prisoner opened dark brown eyes that darted around, but the man didn't appear fully conscious. He reached toward the woman, and groaned, "Amel … ia."

Marie leaped to the next person, pulling off the helmet. *Be*

John. Please be John. Her fingers fumbled, pulling off the hood that gripped his face.

He wasn't breathing.

"John!" Marie yelled, lifting his head to her face.

She set him down on the staging platform, gave fifteen chest compressions and two rescue breaths.

"Defibrillator," Malcom said, opening a panel on the wall and tossing the electrodes to Marie.

Marie ripped the suit off John's chest and stuck the device to his skin.

"Clear!" a computerized voice said.

Her husband's body lurched, but remained unconscious.

"Clear!" said the computer.

The body lunched again. And again. And again.

33

I awoke with a start, feeling disoriented. *What ... where?*

Eyes gazed into mine. Familiar eyes. Eyes I'd loved for so long. Marie tore the pads from my chest and wrapped her arms hard around me. We sat, embracing on the floor, rocking back and forth in tears.

"Don't ever let me go," she sobbed.

"My God, you were dead," I said. "We had a funeral and everything."

"Alright, love birds," said the man with her, now back in the driver's seat. "I guarantee the CDF knows what we've done, and they'll send rovers from the base."

"This is Malcom," Marie said. I nodded my head as Malcom backed the *Mount Everest* out. Gunning the accelerator, he returned to the airlock.

"If the CDF is waiting on the other side," I said, "they'll shoot us as soon as we open the interior door."

"You can bet on it. They just killed my partner," Malcom said, staring at us in a rear-view mirror. "Who are your friends?"

"Amelia, Kevin, Avro," I said, pointing at each of my friends in turn. They were in various stages of consciousness, blinking and gasping.

Avro got it together first, and asked, "Is that a single chamber airlock?"

Malcom nodded.

"Tell them to evacuate the factory."

"They already did. It's CDF personnel only, in there," Malcom said, a sad smile crossing his face. "I *want* them in there while we do this."

Malcom flipped up a seat cushion and started handing out emergency pressure suits.

From the west, two CDF rovers came into sight, winding their way down a path carved into the slag.

"Here they come," Avro said. "I'd say we've got two minutes."

We pulled up to the open airlock, driving the rover inside.

"Alright everyone out," Malcom said, sprinting to the rear of the rover. He hit the release on the rear hatch, and grabbed a hammer from the rover's tool kit. Turning to Kevin, he said, "Indian guy, grab that saw."

Kevin scowled and picked up the tool.

"Pelé," Malcom said pointing to Avro, "drill, Phillip's head."

Malcom swung himself around the side of the vehicle. Stepping up on the rover's metal tire, he covered the airlock's camera with pressure-foam. He tossed the canister to me and I did the same for the window on the hatch.

"The airlock is mechanical," Malcom said, taking the saw from Kevin. "I'm going to sever the line that prevents both doors from opening at the same time." He sawed off a mechanical arm on the left side of the airlock.

"Drill," he said to Avro. Malcom took the drill, and unscrewed a panel on the wall. He ripped out a disk, and passed it to Kevin. "Pressure sensor," he said. Several wires protruded from the sensor. Kevin passed Malcom a pair of plyers, which he used to clip the wires and then splice two of them together.

"Hurry," Amelia said. "The other rovers are right behind us!"

"Just like hotwiring a car way back in the good ol' days," Malcom said. "Ya'll will want to stand clear."

We exited the airlock, back onto Callisto's surface, and hid behind the slag. The approaching rovers were less than one hundred meters away. Malcom went back into our rover, ripping off its sun visor. He joined us behind the mound.

When Malcom hit a button on the transmitter, air shot from vents in the side of the airlock, and a light flashed green. "Opening interior door," he announced.

The door rolled open. As soon as there was a gap wide enough for a man, the vacuum of space snatched its first victim. A single CFD solider shot out, cartwheeling in a rush of air. The opening widened and another soldier flew by, then another. The soldiers hit the ground, bouncing back into the air, guns and grenades falling from their ragdoll bodies.

"Oh God!" Marie cried. "They're not wearing spacesuits!"

"An occupational hazard," Amelia replied.

I hugged Marie against me, wishing I could shield her from the horror and pain.

The door opened all the way and an entire platoon was sucked out through the opening. A dozen soldiers hit our rover on the way out.

Our rover began sliding backwards, and then accelerated, literally flying out of the airlock. It flipped over, tumbling down the path and colliding with the leading CFD rovers at ninety-six kilometers an hour. Glass, warped steel, and tires shot into the vacuum, before arcing back to the surface.

The second CDF rover drove onto the slag to avoid the carnage. It off-roaded for several meters, then returned to the path. More soldiers flew from the open airlock, guns and other gear flying after them. The rate of exiting bodies began to slow; they were like the last popcorn kernels in a microwave.

"Shutting interior door," Malcom said, after it appeared that no one else was coming through. If there was anyone else near the manufacturing facility, they were probably running for their lives.

The ejection of air ceased. Around us lay dozens of dead soldiers, all with red eyes, and frozen blood around their nostrils. The near vacuum of the surface had killed them before they hit the ground.

Avro ran to the nearest solider, plucking a grenade from his belt. He pulled the pin and lobbed it at the surviving rover. He repeated the process again, with another two grenades. One grenade landed under the rover, while the other wedged itself between a hydrogen tank and the hull.

"Get down!" he yelled. We dove behind the nearest slag bank. The grenades went off, and the CFD rover exploded, fire leaping from the shattered windshield, and exterior hatch, but was quickly extinguished by lack of oxygen.

We came out from behind the pile of rocks when the coast was clear. Marie held back, clearly traumatized. I put an arm around her, guiding her toward the airlock.

"Collect their weapons and ammo," Avro said.

We picked up CDF rifles, slinging them around our backs. Kevin filled a satchel with magazines and C4 breacher charges.

Marie leaned against the airlock's interior while I filled my VR helmet with grenades before heading into the airlock myself. Once inside, Malcom and Kevin manually slid shut the exterior airlock door.

Avro and Amelia clinked their weapons together, as though in a toast before a meal. Avro put an arm around her.

Marie looked at me through her spacesuit's visor. "Who are these people?"

I smiled. "They're my best friends."

⚡

With the room repressurized, we stripped off our spacesuits. Several bodies lay on the airlock's floor.

Avro, Kevin, Amelia, and I wore nothing but the resistor suit liners. We pulled tunics, pants, and shoes off several fallen CDF soldiers.

"I'm sorry you have to see this," I said.

Marie sighed. "I'm starting to get used to being traumatized. I witnessed the end of the world, remember?"

"Here, take this," Amelia said, handing a rifle to Marie. "They say shooting is a good stress reliever." Marie took the gun with some hesitation, and slung it awkwardly over her back.

The factory floor outside the airlock was clear of people. Depressurization had severely damaged the facility. The factory roof had collapsed in several places, and the southern wall was gone.

"I assume this is one of yours?" Malcom said, pointing into an adjacent airlock. "They brought it in here to install a new gun."

Avro and I peered at the machine inside. It was one of our JJ's, with the nuclear reactor missing from its rear.

"That's one of ours alright," I said. "But where are the others?"

"They're being held at the base; apparently they've had pilots training in them," Malcom said. "I'm going to leave you now. This place is a mess and I doubt they'll figure I helped you. I have a wife and two kids now and I'd really like to see them again."

"Thanks, Malcom, for everything," Marie said. Malcom nodded, and walked away, stepping over the debris that had accumulated against the airlock door.

"What now?" I said. "Do we go after H3?"

"Not this time, Johnny," Avro said. "Our mission is over; we need to leave Callisto, and get a message to Earth."

"How?" I said, pointing at the airlocks containing our spacecraft. "These things are paperweights. Without the reactor's they can't escape the system. What are we supposed to do, fly just far enough away to get a signal out, then wait in orbit to get rescued?"

"If we have to," Avro said.

Marie cleared her throat. "H3 has a saying: 'always have an ace in the hole'."

We all looked at her.

"You sound like you have a plan?" I said.

"H3 has a cabin in the woods about sixteen kilometers west of Clydesdale. That's where I found him. Behind his cabin, on the other side of the Ring wall, is his spaceship. If I had to guess, I'd say it's fueled and ready to go."

A smile ran across all four of our faces.

"The spaceship is probably locked, right?" Amelia said.

"Everything will be locked," I said. "But we'll figure it out. We always do, right Kevin?"

"Getting inside is easy," Kevin said, reaching into a satchel and pulling out a square of C4. "You're asking me if I can hotwire a spaceship."

"That's exactly what we did, like three months ago," I said.

Kevin just looked at me then smiled. "I am sure we will figure something out." Kevin held up the visor from his VR suit, and a watch from a CDF soldier. He entered the airlock, and synced his watch to the JJ before returning a few seconds later. "Got all my files back," he said, looking at the watch.

"They might come after us out there," Avro said. "We'll need someone to clear the space."

"What do you have in mind?" Amelia said,

Avro looked over at the spacecraft. "You think the ship will remember me?"

Kevin nodded toward the ship. Avro walked over, the ship scanned his face, and the hatch dropped open. He peered inside, and ran a quick diagnostics check.

"Get H3's ship in the air, Kevin, and I'll cover you," Avro said. "Amelia, I love you."

She ran over, jumping into his arms, and gave him a big kiss.

"We'll dock with you in orbit," I said, exchanging a fist bump with Avro. "Good luck, brother, see you in a couple hours." Avro hit the door panel on the JJ, closing it behind him while Kevin cycled the airlock.

We made our way to the exit, stepping around overturned machines and pieces of the structure's roof.

Outside, a Depress-Department truck pulled up to the curb. Seven men in spacesuits poured out, and ran toward the factory like fire fighters to a burning building. Two of the men in tandem carried a tube of pressure canvas.

"Hey," Amelia yelled. They didn't hear her beneath their helmets. "Hey!" she yelled again, grabbing one man by the arm. He stopped, popping the seal on his helmet, and tucked it under his arm. He ran a hand through his brown hair.

"The airlock is secure," Amelia said. "But we lost three platoons."

"What happened?" the brown-haired man said.

"Bastards blew the airlock during the escape attempt," Amelia said.

"Attempt?"

"The prisoners were killed."

"Are you sure?" the man said.

"Saw it myself. They were in the airlock, in a rover. But when the doors blew, the rover went with it. Bodies are all over the surface."

"Thanks, we'll send an EVA team to investigate." The man put his helmet back on, and tapped the transmitter on his wrist.

Several CFD vehicles were parked outside, their owners lying dead on the slag. I walked over to a jeep, reached in, and hit the ignition. Hydrazine engines hummed to life as water vapor poured from the exhaust.

A message crackled over the jeep's radio: "CDF forces, CDF forces, situation secure. I repeat situation secure, return to standard patrols."

"Seems like we've bought ourselves some time," Amelia said.

Marie put a finger on her watch and voice texted Charles. "Something important has come up. Get Branson and Lise, meet us at the Clydesdale stables. I'll be there soon."

⚡

We left Newport, passing a few other patrolling CDF vehicles. They waved as we passed. Our disguises were working, at least for now.

Kevin climbed into the bed of the truck.

"What are you doing?" I asked.

Kevin wore his visor, and was typing on invisible displays. He lifted the visor, and began fiddling with blocks of C4 and a coil of wires.

"You're building a key," I said.

"Of sorts." Kevin shot me a tired smile and held up the C4. I left him to his work.

We reached Clydesdale as the holographic sun perched low on the horizon. A breeze blew from the east and shadows danced on the houses and barns. We passed a wooden wharf with several boats rocking in the subtle swell.

"Marie!" came a man's voice from nearby. We turned to see a blue house with a tall sloped roof. Two dormers protruded above a wraparound deck. The man rushed over from the deck, his eyes red and wet with tears.

"Oh, no!" Marie said, and Amelia cocked her rifle, but kept it hidden below the dash. "Stop the jeep."

Marie got out and the man rushed over, placing his hands on Marie's shoulders. "I got a call from Hoshi; she said you were dead!"

I could tell Marie and this man had something special between them, and deep down, I felt jealous. *How is a man is supposed to feel in a situation like this?* I thought, and suppressed the emotion as best I could.

Marie studied the man's face. "James, you didn't know?"

"I'm so happy you're alive," the man called James said, he looked at us. "Thank you, thank you. I'm so relieved she was wrong ... I ..."

"You didn't know," Marie said again, "about Earth?"

"Hoshi said you were killed in an escape attempt with the Alliance."

"James, there is no Alliance," Marie said. "There never was. These are my friends, from Earth."

"Oh my God. Oh my God!" James said, scanning each of us. Then he gazed at the horizon. I could tell he suddenly saw this colony in a whole new light. "That means my father … dammit, what an idiot I've been."

"Your father had something to do with this?" Amelia said. "If he's still alive, he's in a hell of a lot of trouble."

Marie reached up and hugged James. "I forgive you for lying about the *Klondike*," she said. "I understand now why you did. We're leaving Callisto. These people are from NASA, but it's not safe for us here anymore. They believe we can commandeer H3's ship."

"If there's anything you can do to throw the CDF off our trail, that would be greatly appreciated," I said.

James nodded. "CDF jeeps have been driving by all day. You were never here."

"Thanks," Marie said.

"I told the truth when I said I'd always be there for you, Marie."

Marie nodded, and got back in the jeep. "Good bye, James."

I looked at James, then at Marie. "Nice town," was all I could bring myself to say as we pulled away, leaving James alone.

"That was my home," Marie said, pointing to a yellow Cape Cod. I gave an impressed nod. "Branson and Lise are waiting for us at the stables with Charles, an old friend of mine."

Another old friend? It was going to take me time to adjust to the fact that she and I had been leading separate lives for some time now.

"You said H3's cabin was sixteen kilometers east of here," Amelia said. "This was the last town, when we flew over, and I don't recall seeing any roads."

"There are no roads," Marie said.

"Does your friend have a boat?" Amelia asked.

"I am not getting on another boat," Kevin said. He was wearing the VR headset while programming the stolen watch.

"No boats," Marie said. "Horses."

Amelia and I looked at each other, then back at Kevin. "Horses will do," he said.

We pulled into a dirt path between two houses. The path led to a barn surrounded by a white fence.

Marie reached her arm out the window and waved. A man came out; he looked older, in his sixties perhaps, with a grey beard accented with white. He wore a flannel shirt and looked as if he belonged at the barn. The man unlocked a gate and let the jeep pass.

A young boy stood at the barn door. He was about one meter tall with wavy hair. He looked confused as I got out and walked toward him, tears pooling at my eyes, the gravity too weak to drag them down my cheeks. I knelt in front of him, looking up into his brown eyes. "Branson," I said.

"I know you from somewhere," he said. "Mom? Who is this?"

Branson came closer, studying my face.

Marie was sobbing through a smile.

"Mom, this looks like Dad, like the man from the pictures." Branson was articulating like a boy wise beyond his years.

A toddler dragged a stick across the barn floor. Lise! I smiled and the girl toddled away, hiding behind a feed bucket. Lise looked exactly like the pictures I'd seen of Marie as a child. The toddler's curly brown hair parted around the cutest silhouette of a face I'd ever seen. Marie walked over and swept Lise into her arms to kiss her. The others stood by in silent observation.

I put a hand on Branson's shoulder and smiled. "Branson, I'm your father. We're going to leave this place."

"Marie," the grey-haired man said. "What's going on?"

"They're from Earth, Charles," Marie said. "And Earth is fine."

"Well, technically," Kevin said, "we came from the moon, via Mars, via Earth, but ah, yeah, Earth is fine. California is messed up though."

Charles stood, inspecting us. "They said you were from the Alliance."

"Yeah," Amelia said. "We heard that."

"We need to leave Callisto," I said. "The world, Earth, needs to know what happened here."

"Well …" Charles said, trailing off. He looked up at the sky, then at the south wall, then the north. "I have to say, I like what they've done with the place, even if I disagree with how we got here. I look forward to living out the rest of my days here."

"I'll miss you, Charles," Marie said, giving him a hug. "Tell Diana, thanks. For everything."

"I've saddled up three horses," Charles said.

"We'll need to saddle up one more," Marie said. "I'm sure your wife won't mind if we borrow Jags."

Charles smiled. "He'll find his way back."

Marie and Charles saddled Jags while Amelia and I disassembled the CDF jeep and fed it into the barn's recycolizer, along with the day's muck from the stalls.

34

The horses' motions were fluid and graceful, a unique gait in Callisto's low gravity. We trotted single file down a beaten path behind the town, a trail known only to Marie and a few of the other riders in Clydesdale. Lise hung in a special equestrian harness on Marie's front; by the look of it, she was used to the experience. Branson wore a black helmet and rode with me.

Branches brushed against my tunic and I ducked to avoid others. Along the trail, we occasionally got a glimpse of the Ring's wall, but as darkness covered the Ring, we saw nothing but trees.

The trail ended at the lake which reflected Jupiter's light.

"We're almost there," Marie said.

Marie pulled on her reins and her horse halted. Our horses followed their leader, and slowed to a stop as well.

"It looks abandoned," Amelia said.

"Amelia, Kevin, follow me," I said, dismounting and unslinging the rifle from my back. "But keep quiet."

The cabin was locked, as we had expected. Kevin blew the door open with a small chunk of C4. Amelia stood watch and Kevin searched the interior. I went back to help Marie with the kids, and to release the horses. Kevin flicked the lights on and off, giving the all-clear, and we went inside. Marie set Lise down on a chair, and pointed up a spiral staircase.

"The ship is up there," she said.

Kevin and I climbed the stairs while Amelia kept watch at the door.

"Take this," Kevin said, handing me the stolen CDF watch. "Sync it to the ship's computer once you're inside. Make sure you do it. I've loaded it with my profile from the JJ."

"Why can't you do it?" I said.

"I'll be busy."

"How's it coming with that door?" Amelia called up the stairs.

Kevin put a hand on my shoulder. "There's something I have to do," he said, and then descended the stairs and walked out the cabin door.

"What?" I said. "Where are you going?"

"To get the key to the spaceship."

"Oh, no." I flew down the stairs after him. Amelia came to the door and we looked at each other, trying to figure out what was going on with Kevin.

"Shit," Amelia said. "Grab your gun."

"What's going on?" Marie said, rocking the toddler in her arms.

"Kevin called H3," Amelia said. "He must have known we wouldn't be able to take off without him."

Kevin reached the dock first, standing there with his hands in the air. A boat pulled up to the shore and two CDF soldiers jumped out, grabbed him, and forced his hands behind his back.

Amelia and I took cover behind a tree. I nodded and we moved to the next tree, maintaining our cover. A third CDF soldier stood at the helm, behind a bulletproof shield.

I peered around the tree, and watched as they dragged Kevin toward the boat.

"One word from either of you and I'll blow his head off," echoed a voice from the boat. H3 stood on the deck, behind a bulletproof shield, pressing the transmit button on a microphone. A loudspeaker on the boat's roof amplified his voice.

I was about to yell, but Amelia slapped a hand over my mouth. "He can't let us talk, or the CDF will know we're not from the Alliance," she whispered.

"So, what?" I whispered.

"So, we'd all die, that's so what. How the hell does Kevin think he's going to get out of this?"

"I don't know, but I'm sure he has a plan."

Kevin stood behind the glass now. Two guards pointed rifles at his side.

H3 put the microphone to his lips. "At least one of you has some sense," he said. "Patel here has agreed to complete my spaceships in exchange for Marie's freedom. Amelia will remain as collateral, and Mr. Orville will go back to work. I believe this is an acceptable deal, considering the circumstances."

"Considering we just took out a third of his army, I'd say that's pretty generous," Amelia whispered.

"How the heck is he going to keep his promise?" I whispered.

"You're probably wondering how I can keep my generous promise?" H3 said through the megaphone. "You know I won't kill you, John. I need you. We need you to finish what you started. Amelia, however, is definitely expendable at this point. So, if you want to keep your friends alive, it's best we all cooperate."

"Thanks," Amelia whispered.

"Here's how it's going to work," H3 said. "John, come with me and we'll unlock the ship, and send it on its way. Amelia will continue her cute little standoff. Any deviation from the plan, and my team has orders to kill Mr. Patel here, and blow up the cabin."

H3 got out of the boat and began walking up to me and Amelia. I noticed he was wearing CDF armor, a bulletproof vest, and helmet. In one hand, he carried a side arm. As he got closer, I clocked my gun, so angry I was still debating whether to shoot him

He got closer, and called me out: "John, come with me, leave your weapon."

I looked at Amelia, and handed her my gun.

I walked toward H3. I could tell he was scared, his secret hanging by a thread. He raised the gun to my face. For a moment, I thought he was going to shoot. I could tell he hated me with everything in his being, but this was a standoff. If he fired, it wouldn't end well for anyone.

"You son of a bitch," Amelia whispered, just loud enough so that he could hear.

H3 took another step forward, positioning the gun only centimeters from my face.

Then I heard it. A voice from the boat. "Good bye, my friends."

The soldiers looked at H3, waiting for orders, not sure if they should shoot.

H3 was infuriated, and looked down toward his guards. "Shoot him!" he yelled.

But Kevin hit a trigger he'd been hiding in his belt, and the C4 he'd secretly hidden beneath his tunic exploded. In one thousandth of a second, Kevin was gone.

The boat's hydrogen tanks ignited in the blast, and the lake was illuminated like day. I could see the whites of H3's blue eyes, and in them, I saw the realization that he felt truly alone. The shockwave from the explosion almost knocked me off my feat, but the blast had disoriented H3 as well.

H3 began to regain his composure, and as the pistol approached my forehead, he said, "Listen here, you sack of shit." The gun was firmly pointed between my eyes. "There's only one person calling the shots around-"

My hand rose to meet it, pushing the gun to the left, tilting my head to the right. Amelia screamed. H3 pulled the trigger, the gun fired and the bullet struck a nearby tree. I yanked his arm down and around, pivoting my body, twisting the gun out of his hand, breaking his fingers. He groaned. "That's for Kevin." I kicked his leg from the left, taking him out at the knee. I wound his arm behind his back and broke it at the elbow, and H3 screamed in agony. "That's for threatening my family."

Amelia ran over, ripped off H3's helmet, and pressed her rifle to his head.

"Will the spacecraft read the bios off a dead man?" she said. It was an idle threat.

I wrestled the fugitive to his feet, and began dragging him toward the cabin.

"We can make a deal," H3 said, holding his broken arm against his chest. "I'll let you leave, just leave me here on Callisto."

"You have no leverage, old man." Amelia pressed the rifle into his back.

We approached the cabin. I realized Marie had shut off all the lights; she was probably terrified.

"Marie, it's us!" I yelled as we approached the cabin. "It's all clear."

We opened the door, and Amelia grabbed H3, forcing him up the stairs. As H3 approached the airlock, it registered his presence and hissed open.

I found Marie and the kids, hiding in the bathroom.

"I heard an explosion, what happened?" Marie said.

"I'll tell you once we're onboard," I said, realizing I had tears in my eyes.

Marie carried Lise while I grabbed Branson's hand and led

the way to the ship.

With H3 on board, the ship came to life. “Welcome back, Henry,” the ship’s computer said.

Marie’s face contracted with anger when she saw H3. She rocked back, ready to throw a punch.

Amelia grabbed her arm. “There’ll be time for that later.”

“He is not coming with us,” Marie said.

“Nope,” I said. “He’s not. Just until we reach orbit over Callisto.”

“Where’s Kevin?” Marie said, looking around.

“He’s …” I looked at Branson, who was strapping himself into a chair, “gone.”

The interior of the spacecraft was nearly identical to the one we had taken from Mars on our trip to the moon. With the ship under gravity, chairs with seat restraints had risen from the floor, ready for launch.

I took Kevin’s stolen CDF watch out of my pocket and pressed it to the consol.

The computer came online, generating a new authorization profile as Kevin’s program cycled through a list of commands.

“Okay,” I said, reading off a display near the front of the craft. “The computer has transferred control from H3. Marie, get the kids secure,” I ordered. “I’m going in.”

Two VR crèches located near the front of the spacecraft served as the cockpit. I threw myself down into the right crèche. Zippers self-climbed from my feet to my neck, securing my body into the resistance suit. I pulled down the face mask, and immersed myself in the cockpit.

“Amelia, this thing has a gun, right?” I said.

Amelia flipped through a menu on the console. “Affirmative. Someone has modified the asteroid guns to serve as defensive weapons.”

"Outside!" Marie yelled, as a fighter-plane-like spacecraft streaked past.

Then another identical craft flew by in pursuit. Guns blazed from the pursuing craft as the leading spacecraft blew to pieces.

I activated the ship's short range optical wavelength transmission system. "Orville here. Avro, do you read me?"

"Avro here, five by five."

Avro's JJ flew past the starboard side of the cruiser, tipping its wing in the classic pilot's hello.

"Good to see you, brother," I said. I was about to tell him about Kevin, but couldn't bring myself to yet.

"Everyone secure back there?" I said.

Amelia was busy duct taping H3 to a seat. "Almost done here," she said. Amelia had wrapped a piece of tape around H3's mouth. Confident H3 was secure, she climbed into the AR crèche beside me.

"Get us out of here," Marie said.

"Yes, ma'am. Retracting the hatch," I said. The ship rumbled as the systems came online. We heard the mechanical creak of the retracting jet way.

"Temperature and pressure nominal," I read off a checklist. "Cryogenic tanks at one hundred degrees Kalvin. We've got to cool the tanks to ten degrees before trans-solar injection. The computer says that'll take at least ... thirty minutes."

"Shit, that long?" Amelia said.

"We're talking about a pretty long engine burn here. It's not only Callisto's gravity we're escaping from, it's Jupiter's."

"Just get us off the surface, would you?" Amelia said.

I pushed a virtual control column forward and blue flames emerged from the bottom of the spacecraft. The acceleration pressed us into our seats as the cruiser lifted off. Augmented reality fully replaced my view of the ship's interior, granting me a 360-degree view of the outside.

Avro fell into formation with the cruiser. He'd adjusted his AR permissions to allow me to see into his cockpit. "There." He pointed. "Head to those mountains while we wait for the cryo tanks."

"Roger that," I said. "Heading toward the mountains."

Amelia had swung her seat around, scanning the surroundings. "We're about to have company!" she yelled. "Bogeys in pursuit, it looks like more of our JJs."

"Shit," I yelled, my senses heightened by a renewed desire to protect my family.

"I'm on it," Avro said. "Dropping back." He banked away, thrusters glowing against the darkness of the moon.

"Go get 'em," I said.

H3 moaned from behind the duct tape. His eyes bulged and his head twisted from side to side.

Marie sat in a seat next to H3. "Wait," she said. She unlatched her harness, and reached over to H3, tearing the duct tape from his mouth. H3 winced in pain, as the tape ripped his skin.

"Tell them to back off," Marie said. "Tell them you are onboard, and not to attack us!"

H3 swore then said, "I instructed them to destroy this ship as soon as it took off."

"Get on the god damned radio!" I yelled.

"The channel is open," Amelia said.

"It's no use," H3 said, "they're not listening."

"Try!" I yelled.

"CDF Forces, this is Henry Allen the Third, do not shoot at the cruiser, I am onboard, I repeat do not shoot!" H3 looked at me. "It's no use, John. They have orders to ignore any transmissions from this vessel."

"Do you have any suggestions?" I said.

"None that matter," H3 said. "You've killed us all."

"Put that tape back," Amelia said. Marie grabbed the roll from the seat pocket, extracted a fresh stretch of duct tape, and wrapped it twice around H3's head, before strapping herself back in.

Once in the mountains, I banked around a tall peak. I swung my head to keep one eye on the approaching bogeys.

"I see their transponder signals," I said. "Can they see us?"

"Affirmative," said a familiar voice with a distinct Punjabi accent.

Anyone looking at my face would have seen me go white, the blood draining away.

The voice was Kevin's.

"The JJs have formed a virtual network," said Kevin's voice over the ship's internal speakers. "We'll need to break the connection if we're going to have any success hiding in those mountains."

"Amelia, what the hell is happening?" I said.

"I … I don't know," Amelia replied.

"I am Kevin's Turing avatar," said the voice. "Kevin knew you couldn't live without him."

"Kevin's Turing, huh?" I said. "Alright KT, how can you help us?"

"I have silently activated the enemy's JJ training profile," Kevin's Turing said. "We will use that to our advantage."

"Okay," I said. "Can you help me get a lock on the bogeys?"

"Both this ship and our JJ's are outfitted with superior electronic countermeasures. Weapons must be aimed manually, visually, and at close range to be of any affect."

"Then it's going to be a fair fight," Amelia said.

"If you consider we're outnumbered, six to two, then yeah, that sounds fair," I said.

I piloted H3's spaceship between two mountains. Avro banked out of sight, taking another route.

"They're closing on us," Amelia said.

"Roger," I said, banking around the crest of a hill.

One of the JJ's fired, and a rock ledge exploded at my three o'clock. There was no way our ship could take a single hit from those guns.

"Dammit," I yelled. *So much for trying to hide.* "Amelia, start shooting!"

Amelia unloaded several rounds on the attackers, but the shots missed, as the targets were several miles behind us, and too small to hit directly. However, several of her projectiles impacted with the mountain peaks, sending a shower of dust and debris in our wake. The pursuing JJs changed course to avoid it.

"KT, can you operate the other gun?" Amelia yelled.

"I would, but I am busy right now," KT said.

"Doing what?" I yelled, wondering what this Turing program could possibly be working on.

"Shopping," Kevin's Turing replied.

A moment later, another JJ appeared, rising from a nearby canyon. Its lights were blinking, sending Morse code, a series of dot and dashes that I couldn't help but read.

S-c-h-r-o-d-i-n-g-e-r,

"You're such a nerd!" I said.

"What the hell?" Avro yelled over the comm.

"The ship is not real, Avro," Kevin's Turing said over the inter-ship channel. "I've silently activated the Jupiter Jumper flight training profile; and the ship belongs to their Turing instructor, me."

"Good work, Kevin, get in position to draw their fire," Avro said.

I turned to Amelia. "Should we tell him?" I asked. "About Kevin …?"

"Wait until this is over,"

Avro banked his JJ around a mountain, flanking one of the pursuing JJ's. He had a shot, and took it. The pursuing JJ exploded;

the bolts from Avro's gun cut through the spacecraft as if it wasn't even there. Avro pulled up to avoid the debris.

"Kevin, can you make another bogey?"

"Negative," Kevin said. "Any new spacecraft will violate the Schrödinger protocol."

I looked to port. Two CDF JJ's flew parallel to us about a mile out.

"Amelia, do you see …" I said as the enemy spacecraft banked towards us.

"I see 'em," Amelia replied. "Take cover behind that ridge."

"Roger that."

The pursuing JJ's approached fast, and Amelia held her fire. I activated the thrust reversers, a feature designed for runway landings. The spacecraft ground to a halt, hovering just beyond the ridge.

Amelia fired, aiming at the crest. The tiny projectiles left our spacecraft at nearly the speed of light. We jolted to starboard, a byproduct of Newton's second law.

The ridge exploded, and rocks shot into space at deadly speed, creating a cloud of flak, like in the skies over Berlin in World War Two. The pursuing JJ's hit the debris at several thousand miles per hour, and exploded.

"Got 'em!" Amelia yelled.

"Nice work. Only three left," I said, accelerating out from behind the ridge. "We're running low on thruster fuel, I'll need the rest to achieve orbit of Callisto. Avro, can you cover us?"

"Roger, covering."

I pulled up, hitting the autopilot, and double checking the navigation aids as we raced for orbit where, if all went well, we'd dock with Avro, and then initiate the eleven-minute burn to exit the Jovian system.

I pulled off the visor and looked back at Marie. She had Lise secured to her chest, and I was surprised the toddler wasn't crying. Branson sat in his chair wide-eyed. Whatever he'd thought his dad would be like, this probably wasn't it.

Through the windows, we could see Callisto falling away.

I slipped back into augmented reality. "We're sitting ducks up here," I said.

"One bogey, headed your way," Avro said. "Kevin, time to draw them out."

"Roger that." Kevin exited the canyon, taking pursuit of the lead bogey as it approached our ship. He twisted his virtual JJ left and right, knowing they were going to take him out, but staying alive as long as possible, giving Avro time to plan his attack.

The two other enemy JJs came in from the side, flanking the Turing's craft with gun's blazing. The virtual spacecraft exploded. Inside each of the pursuing enemy spacecraft, the CDF pilots would have been notified of the deception, informed that they'd just killed a decoy.

It was the distraction Avro needed. He pulled around, taking out one of the flanking JJ's, then the other. The ships exploded in flames, arcing down to the surface before impacting, creating new craters on Callisto.

Only one bogey left now. Avro pulled up, and with the pursuing JJ only a few miles behind us, he took the shot.

"That's all of them!" Avro yelled.

"Yes!" I yelled, and tossed off my headset.

Kevin's face hovered on the forward display.

"We should tell him," I said

Amelia looked at me and sighed. "Wait until he's inside," she said. "KT, please deactivate the Turing."

Avro docked on the spacecraft's belly, and floated out into the cabin. Amelia grabbed him, and they spun around in a zero G embrace. Lise was crying, her body unused to zero gravity. Marie floated the kids into one of the staterooms, turning on the entertainment system and waiting for Lise to calm down before joining us back out in the main cabin.

Avro took off his VR hood, and looked around. "Where's Kevin? What's H3 doing here?"

Tears welled in Amelia's eyes. "Kevin's dead."

"I don't understand," Avro said. "His ship wasn't real?"

"His ship wasn't," I said. "And neither was he. Kevin saved us, by bringing us H3, and leaving his Turing avatar to unlock the ship."

Avro turned to H3, looked at me. I hit a button on the console, and Kevin's face appeared on the screen.

Avro turned to the screen. "Kevin wanted me to say good bye," said the Turing. "He also wanted me to tell you to knock H3 out, put him in your JJ, and send him into a high radiation, low orbit around Jupiter."

H3 groaned behind his duct tape gag. Avro turned, and slugged him in the face. The punch was so hard it knocked him unconscious.

"I've programmed the orbit into the JJ," Kevin said. "And deactivated the radio. He won't be found until we want him to be found."

"I guess if they want him, NASA can send another team to pick him up," I said.

We tossed H3's unconscious body into the JJ's spherical cockpit. Avro grabbed a rifle, and went inside, using the butt of the gun to smash every computer interface. He came out, and sealed the door, leaving H3 alone inside without an AR suit.

The JJ dropped away, its thrusters fired, and it sailed off toward the gas giant.

"Good riddance to bad rubbish," Amelia quoted.

"And dead weight," I added.

"Will he die?" Marie said.

"Unlikely," Kevin's Turing said. "The ship's auto-chemo machine will treat his radiation sickness as fast as the tumors grow, and there's plenty of supplement and water."

The four of us floated near Kevin's image. It was time to say farewell.

"Goodbye, friend," Avro said, glancing at me and Amelia. We nodded, and Avro keyed in the commands to deactivate the avatar.

Marie noticed the tears my eye and placed her arm around my back. "My friend Lise has a saying I'll never forget. It's the most profound thing I've ever heard, and I think it applies to your friend."

Marie paused, taking a breath before adapting the proverb for Kevin. "When Kevin was born, he cried and the world rejoiced. He lived such that now, as the world cries, he'll rejoice."

Branson came out from the stateroom and floated toward me. For the first time, I realized how lanky he was, just like I was at his age. He wore a huge smile, as he pressed off one wall, doing a summersault and landing on the other.

"You know we're going to have to stick you in a resistor suit," I said. "You've been on a low gravity world since you were two, and gravity's going to hurt."

"Ah, Dad! C'mon!"

"VR is a lot of fun. I've got all sorts of sports to teach you."

"I'm worried, John," Marie said. "What if they can't handle Earth's gravity?"

"We're not going to Earth," I replied.

"Wait, what?" Marie said.

Amelia looked over at her. “We’re going to Mars.”

“The transition to Mars’s gravity will be much easier on the kids,” I said. “After they adjust to Mars’s gravity, and we’ve enrolled them in some high G therapy, then we’ll think about taking them back to Earth.”

“Mars! Is that safe?” Marie said.

“It better be safe,” Amelia said. “Last time I checked, your husband was still the president.”

The End

www.ingramcontent.com/pod-product-compliance
Lightning Source LLC
Chambersburg PA
CBHW021027210326
41598CB00016B/937
9780999034682